BestMasters

Stefan Nanz

Toroidal Multipole Moments in Classical Electrodynamics

An Analysis of their Emergence and Physical Significance

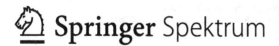 Springer Spektrum

Stefan Nanz
Karlsruhe, Germany

BestMasters
ISBN 978-3-658-12548-6   ISBN 978-3-658-12549-3 (eBook)
DOI 10.1007/978-3-658-12549-3

Library of Congress Control Number: 2015960957

Springer Spektrum

Printed on acid-free paper

Springer Spektrum is a brand of Springer Fachmedien Wiesbaden
Springer Fachmedien Wiesbaden is part of Springer Science+Business Media
(www.springer.com)

Preface

If I have seen further it is by standing on the shoulders of giants.
ISAAC NEWTON

This book is based on my master thesis, on which I worked from January 2014 until February 2015 at the Institute for Theoretical Solid State Physics at Karlsruhe Institute of Technology (KIT). The topic regarding the toroidal multipole moments was originally suggested by my supervisor, Professor Carsten Rockstuhl, who had just started to establish a new working group at KIT at the end of 2013. Pointing out some inconsistencies in various derivations and descriptions of the electromagnetic multipole expansion, he motivated me to dig into the depth of the theory of electrodynamics and to find out what's really going on in the multipole expansion.

Soon after I started working, I had the fortune that the working group was joined by an analytically well-experienced postdoctoral researcher, Dr. Ivan Fernandez-Corbaton, who took the role as my advisor and helped me to tackle the sometimes dry theoretical analysis. In the course of time, it became apparent to us that the derivations which argue in favor of the physical significance of toroidal moments are suffering from some assumptions, which turn out to be not justified in general. Condensing these insights then lead to a master thesis which was approved by the supervisors, Prof. Carsten Rockstuhl and Prof. Martin Wegener, and later on also by Springer Spektrum, resulting in their decision to publish it as a book.

At this point, I want to take the chance to express my cordial thanks to several people without them my master thesis and therefore this book would not have been possible. First of all, I would like to thank Prof. Dr. Carsten Rockstuhl for giving me the possibility to work on such an interesting and rewarding topic and for being a great mentor regarding all aspects of research. I would also like to thank Dr. Ivan Fernandez-Corbaton, who did a great job as advisor with his analytical skills and extensive knowledge of literature.

Furthermore, I would like to thank: Dr. Christoph Menzel for fruitful discussions and proofreading the thesis; Iris Schwenk for providing me a nice template for the layout of the thesis; Dr. Andreas Poenicke for his support regarding computer problems; Clemens Baretzky for creating the images; my half-brother Philipp for proofreading the thesis; and my roommates Dr. Giuseppe Toscano and Alexander Kwiatkowski for sharing the office with me and helping me with all kinds of problems. I am grateful to Springer Spektrum for publishing my master thesis, and I want to thank my contact persons there, Nicole Schweitzer and Marta Schmidt. Last but not least I would like to thank my parents for their support during my education.

Heidelberg,
November 2015

STEFAN NANZ

Contents

List of Figures

List of Symbols

This list includes the most important symbols which are at least used once in the thesis.

- $\vec{\nabla}$ nabla operator
- Δ Laplace operator
- $\mathcal{P}$ parity operator
- $\mathcal{T}$ time inversion operator
- $\mathcal{L}$ angular momentum operator
- $\mathcal{D}$ detracing operator
- Λ detracing functional
- Tr trace of a tensor
- $\vec{k}$ wave vector
- $\dot{x}$ time derivative of x
- a^* complex conjugation of a
- $\vec{r}$ position vector of evaluation point
- $\vec{r}\,'$ position vector of the source
- $\vec{e}_i$ unit vector in direction i
- μ_0 vacuum permeability
- ε_0 vacuum permittivity
- c speed of light in vacuum
- ω angular frequency
- $\vec{A}$ vector potential
- φ electric scalar potential
- $\vec{j}$ electric current density
- ρ electric charge density
- $\vec{B}, \vec{H}$ magnetic field
- $\vec{E}$ electric field

- $\hat{P}^{(n)}$ Cartesian electric multipole moment of order n
- $\hat{M}^{(n)}$ Cartesian magnetic multipole moment of order n
- $\hat{T}^{(n)}$ Cartesian toroidal multipole moment of order n
- $\vec{p}$ electric dipole moment
- $\vec{m}$ magnetic dipole moment
- $\vec{t}$ toroidal dipole moment
- $\hat{Q}^{(e)}$ Cartesian traceless electric quadrupole moment
- $\hat{Q}^{(m)}$ Cartesian traceless symmetric magnetic quadrupole moment
- $\hat{Q}^{(t)}$ Cartesian toroidal quadrupole moment
- Q_{lm} spherical electric multipole moment
- M_{lm} spherical magnetic multipole moment
- T_{lm} spherical toroidal multipole moment
- a_{lm} spherical electric parity multipole moment
- b_{lm} spherical magnetic parity multipole moment
- j_n spherical Bessel function
- $h_n^{(1)}$ spherical Hankel function of first kind

1 Introduction and Overview

The analysis of electromagnetic radiation is an important resource to investigate the properties of materials. Often, unknown materials are irradiated by a suitable light source, e.g. a laser, and the scattered electric and magnetic fields are used to obtain information about the materials. This requires to link the properties of the incident to the scattered radiation. This asks to solve Maxwell's equations with spatially distributed materials whose properties are introduced on phenomenological grounds. A prototypical example for such approach is ellipsometry. When designing e.g. antennas, the opposite is of relevance: An incident and scattered electromagnetic field is given and the structure, which produces this field, shall be constructed.

The latter problem also applies in the context of metamaterials. Metamaterials are made from small scattering objects. If their optical response is dominated not just by an electric dipole moment but also by higher order electromagnetic multipole moments, material properties not available in nature can be reached. This is possible, since on phenomenological grounds natural materials at optical frequencies usually are considered as an ensemble of polarizable entities with a scattering response that corresponds to an electric dipole. This only allows to observe a dispersion in the optical permittivity

Now, by composing metamaterials from strongly scattering unit-cells, also called meta-atoms [1], other material properties are allowed to be dispersive. The key is to make meta-atoms sufficiently small and to arrange them sufficiently densely in space, such that light will experience a homogeneous medium with properties derived from the scattering response of the individual meta-atom. For example, if their scattering response is dominated by a magnetic dipole moment, a dispersive permeability can be observed. If their scattering response is dominated by an electromagnetic coupling, a strong optical activity can be observed. Studying the light propagation in metamaterials and the scattering response from individual meta-atoms is a major challenge for contemporary theoretical optics.

From the many interesting properties meta-atoms may exhibit, we discuss in this thesis the toroidal multipole moments. Toroidal moments are a third multipole family besides the electric and magnetic moments. They have several properties which attract the interest of research. It has been shown [2] that current distributions, which cause toroidal moments, violate Newton's third law. With toroidal moments it is possible to generate non-zero vector potentials without electromagnetic fields [3], hereby realizing non-radiating charge-current distributions [4, 5]. Also, a negative index of refraction can be caused by toroidal moments [6]. Though the experimental possibility to realize materials with such properties is quite new, there has been a theoretical interest for such characteristics for long time. Although being considered in the context of electrodynamics usually as exotic, the properties of toroidal moments came again in the focus of interest with the advent of metamaterials. Particularly the similar field distribution of electric and toroidal moments has been a fascination, but also a source of confusion in the contemporary literature. Disentangling the contributions of electric and toroidal multipole moments is therefore important and may also shed new light on supposedly well-known systems. For example, dielectric spheres [7] and nanowires [8] have recently been reported to have not-negligible toroidal moments. **It is, therefore, the purpose of this thesis to understand the origin of toroidal multipole moments from first principles in the context of a multipole expansion.**

For many decades, such toroidal multipole moments were neglected in the treatises of multipole expansion. It is worth noting that in standard textbooks no toroidal moments are mentioned [9–11], maybe because of their insignificance in experiments at the time when the books were written, or maybe the authors were not aware of them. In most cases, the toroidal moments are treated as a part of the electric multipole moments. This is often an appropriate approach, because the toroidal moments are extremely weak in natural materials and the fields of toroidal moments are similar to those of electric moments. However, in meta-atoms their contribution can get as strong as that of the electric multipole moments. It is then not sufficient anymore to just attribute the toroidal multipole contributions to the electric contributions. This would e.g. leave unexplained vanishing electromagnetic fields due to destructive interference of the fields of electric and toroidal moments. Also, when considering time inversion, the fields show different behavior, meaning that the sum of both fields changes also. This could not be explained with just electric multipole moments, because they do not change under time inversion. One also wishes to have multipole moments that behave well under rotations, meaning that the angular momentum properties remain invariant. This is not the case when the toroidal multipole moments are mixed up with the electric and magnetic multipole moments. For all reasons we wish to clearly state there is a necessity to correctly distinguish the toroidial moments from the electric and magnetic multipole moments.

To enable such distinction, we will investigate in this thesis several approaches to study the emergence of toroidal moments from the basic equations of electrodynamics within the frequently used formalism of multipole expansions. Already in the 1970s and 1980s [12], but also in recent years [13], there were a few publications on quite elaborate and cumbersome formalisms how the toroidal moments can be deduced from general representations of the charge and current densities. However, this is not the focus of this thesis. It is more dedicated to formulate the standard electrodynamic multipole expansion in a way so that on the one side the derivations from textbooks are used, but on the other side the toroidal moments are included. We will show that the usual Cartesian Taylor expansion has a few disadvantages regarding the proper definition of multipole moments, and outline other methods how the multipole moment tensors, including the toroidal ones, can be defined consistently.

Structure of the Thesis

After this introduction in chapter one, we start in chapter two with a brief history of the research about toroidal moments. Furthermore, reasons are given why toroidal moments are needed besides the standard electric and magnetic moments for a complete description of arbitrary charge-current distributions. For this sake, we will discuss several prototypical charge-current distributions and symmetry properties of multipole moments. Also, we give a brief delimitation of toroidal moments and anapoles because they are often mixed up in literature.

In chapter three we introduce the most important equations necessary to perform the electrodynamic multipole expansion. These are Maxwell's equations, the continuity equation and wave equations as well as their solutions. The useful formalism of potentials will also be outlined, introducing the scalar and vector potential and the Debye potentials. We will also clarify our notation and symbols.

In chapter four we present at first briefly the standard way of doing the multipole expansion in Cartesian coordinates based on the potentials. We discuss advantages and disadvantages of the Taylor series and define the multipole moment tensors in two ways. One way ignores the existence of toroidal moments and contains multipole moment tensors without definite properties under parity and rotations. Hereafter, we show how from those inconvenient and inconsistent tensors the correct physical multipole moment tensors, including the toroidal ones, can be derived and how this can be motivated from an exact decomposition of the vector potential into tensors with distinct symmetry properties. Additionally, we discuss a method of how arbitrary multipole moment tensors can be calculated and how the vector potential can be expressed with these tensors. In the last part of the chapter, the multipole expansion is performed in spherical coordinates. We compare this with the expansion in Cartesian coordinates and show how the toroidal multipole moment tensors emerge from a decomposition of the current density in momentum space. This derivation will raise the question if the toroidal moments are a full degree of freedom of a given system, or if they are always related to the electric dipole moments.

In chapter five we perform the multipole expansion based on a decomposition of the electromagnetic field. We will see that with this decomposition, it is not possible to distinguish between electric and toroidal moments. In a second approach, we will additionally use the charge and current distributions to express the expansion coefficients in the field decomposition. We will show that even with the sources, the electric and toroidal moments are still mixed up, but can be disentangled in the long-wavelength limit.

In chapter six we will outline how the transformation between the different multipole moments can be done. We will give transformation formulae to express the Cartesian multipole moments with the spherical moments. We will then discuss the fields of toroidal moments, both in the far field approximation and in the exact form. We will also show how a non-radiating current distribution can be generated using electric and toroidal dipoles.

In chapter seven, finally, a summary of the thesis is given, together with an outlook to possible future experimental realizations. We outline remaining open questions which could not be answered as part of this work.

Notation

The most used symbols are listed with a short definition in the appendix. Basically we use Einstein summation convention, however, in many cases, the summation symbol is denoted redundantly for clarity. A scalar product is always indicated through a dot ".". The imaginary unit is written as i, the index variable as i. ϵ is the Levi-Civita symbol, ε the scalar permittivity. e is Euler's number, the superscript e at tensors indicates the electric origin of these quantities. Tensors are denoted with majuscules and a hat, tensor entries with majuscules and subscripts and operators with majuscules in calligraphic style. φ is the electric scalar potential, whereas ϕ is the polar angle in spherical coordinates. t is the ordinary time and τ is the retarded time.

We further distinguish between "multipoles" and "multipole moments" in the following sense: A multipole is a term in the multipole expansion of the vector or scalar potential or of the fields. A multipole moment is a numeric, in general tensorial quantity, which is characterized through the charge or current distribution. A multipole consists of a multipole moment and a coordinate dependent factor. We use "multipole moment" and "multipole moment tensor" as synonyms, but in the first case, the emphasis is on the physical aspect, whereas in the latter case, the emphasis is on the mathematical aspect.

2 Why another Multipole Family?

This chapter serves as a phenomenological motivation for introducing toroidal moments. With this, it lays the ground on which we will unfold the mathematical details in the following chapters. We start with a short overview regarding the research efforts concerning toroidal moments in the last decades. This past research already provides a considerable understanding of the toroidal moments, but also of the concept of multipole expansion in general. A discussion of several prototypical charge and current distributions that possess specific electromagnetic multipole moments is done in the following. Hereafter, we consider symmetry arguments and fundamental physical concepts to support the idea that there exist three multipole families, i.e. electric, magnetic, and toroidal multipoles. In the last section we discuss the distinction between the toroidal moment and the anapole.

2.1 A Brief History of Toroidal Moments

Toroidal moments have not been examined for a long time in the theory of electromagnetic multipole expansion. In 1957, Zel'dovich discussed parity violation of elementary particles and postulated that spin-$1/2$ Dirac particles must have an "anapole" [14]. In the late 1960s and early 1970s, Dubovik *et al.* [15] connected the quantum description of the anapole to classical electrodynamics by introducing the "polar toroidal multipole moments". The term "toroidal" stems from current distributions in the shape of a circular coil that were first shown to have a toroidal moment. "Polar" indicates that the polar toroidal moment transforms under parity as a polar vector. Dubovik *et al.* showed that the toroidal moments form, like the electric and magnetic multipoles, a family of multipole moments and managed to express the toroidal moments in the language of the classical multipole expansion [12]. Already at that time it was pointed out that for symmetry reasons a fourth multipole family potentially could exist. They were called "axial toroidal moments", because the dipole moment, which belongs to this family, transforms under parity like an axial vector. We will outline this later, but it is worth emphasizing that a distinction is made here between axial and polar toroidal moments. However, we do not encounter the axial toroidal moments due to the non-existence of magnetic charges and currents [16]. In 1986 [17], Dubovik *et al.* discussed the existence of such axial toroidal moments in continuous media and related it to phenomenons like phase-transitions and non-linear optical effects.

Toroidal moments were not acknowledged outside the Soviet Union as being an important part of the multipole expansion until the 1990s and remained neglected from a big part of the research community even in the 1990s and early 2000s. In 1988, Bladel [18] discussed several current distributions that required the concept of toroidal moments for their understanding. He pointed out that another term, besides the well-known electric and magnetic multipole moments, is necessary to properly describe an arbitrary current distribution. He was probably not aware of toroidal moments, since he neither referred to the papers of Dubovik *et al.* nor mentioned that this term belongs to another multipole family. In their 2005 published book [10], the authors Raab and Lange discuss the difference between fields originating from different definitions of the multipole moment tensors, but do not mention a connection to the third multipole family.

In the 1990s, Afanasiev *et al.* published a series of papers where they analyzed in detail the fields of toroidal moments [19] and discussed specific current distributions which produce time-dependent toroidal moments [20]. In 2000, Dubovik *et al.* [21] discussed the properties of continuous media with toroidal moments, introducing the so-called "toroidization", an analogon to polarization and magnetization, which characterizes the effect of a toroidal multipole field.

However, despite this increasing theoretical understanding of toroidal moments, experimental evidence was scarce. The toroidal moments only came into consciousness of the research community with the advent of metamaterials. They are composed out of basic unit cells, called meta-atoms. For meta-atoms with a strong electric dipolar response, strong dispersion in the permittivity is observed. A meta-atom with a strong magnetic dipolar response causes a strong dispersion in the effective permeability. Extrapolating from these insights, it should be possible to identify meta-atoms that have a strong toroidal moment.

Basically with the possibility to design meta-atoms with high toroidal dipole radiation, it became necessary to take this contribution into account. In 2002 [22], Radescu and Vaman published a detailed analysis of radiation properties of arbitrary charge-current distributions including the toroidal moments. In the same year, Vrejoiu [23] proposed an algorithm which makes it possible to calculate the toroidal moments from the usual Cartesian multipole expansion.

In the early 2000s, Zheludev, Fedotov *et al.* began working on various properties of toroidal metamaterials. They carried out lots of simulations and analytic calculations, which improved the understanding of the properties of toroidal moments. In 2007, they calculated the anapole moment of nanocrystals [24]. Turning to possible applications of toroidal moments, they showed in 2009 that a torus-shaped structure provides optical activity [25]. In 2013 and 2014, they analyzed non-radiating structures [26] and transmission and reflection properties of thin layers of metamaterials consisting of toroidal dipoles [27]. Most recently, toroidal moments have been found in dielectric nanoparticles [7] and nanowires [8]. In 2010, Kaelberer *et al.* carried out an experiment measuring toroidal dipoles in a metamaterial slab consisting of split-ring resonators [28].

In the following we provide a condensed summary from the insights of this past research to understand on the grounds of basic considerations regarding charge-current distributions the origin of toroidal multipole moments.

2.2 Characterization of General Charge and Current Distributions

Every multipole moment is associated with a specific charge-current distribution. An arbitrary source distribution can be expressed through a sum of terms, called "multipole expansion". Each term is a "multipole" and consists of a numeric weight factor, called "moment" or "multipole moment", and a functional dependency of the coordinates. Moments are tensorial quantities and enumerated as exponentiation with the base two, thus, the n-th term in the sum is a 2^n-pole. [29]. Each of the multipoles has a different radiation pattern. It turns out that all radiation fields can be classified by using the symmetries of spatial inversion and time reversal. This will be analyzed in the next section.

To illustrate this concept of fundamental charge and current distributions, we discuss now the sources of the four dipoles, i.e. the electric, magnetic, polar toroidal and axial toroidal dipole. They are illustrated in Fig. 2.1. The simplest radiating configuration is given by two opposite and separated charges, Fig. 2.1a. This configuration generates the electric dipole moment $\vec{p}$. Inverting this configuration reverses the direction of the electric dipole moment. This means that the electric dipole moment is antisymmetric under parity transformation. Inverting the time does not change anything for the dipole, because by convention, the electric charge does not change sign under time inversion [9, p. 271].

Another basic current configuration is the circular current, pictured in Fig. 2.1b. This current distribution is divergence-free, as all current lines are closed. Such a configuration causes a magnetic dipole moment $\vec{m}$ that is perpendicular to the plane in which the current flows. The direction of $\vec{m}$ depends on the direction in which the current flows. If the current direction (or equivalently the time) in Fig. 2.1b is reversed, the magnetic dipole moment would point downwards. This means that the magnetic dipole moment is antisymmetric under time inversion. Inverting the space changes nothing for the magnetic dipole moment, since the inversion of the spatial coordinates and of the current direction compensate each other.

The current configuration, which produces a polar toroidal dipole $\vec{t}$, is pictured in Fig. 2.1c. It consists of circular currents which are symmetrically arranged on the surface of a torus. Often, one finds also pictures which show a coil bent into a ring [6]. Like the circular current before, the divergence of this current distribution is zero. Inverting the directions of the current will invert the direction of the toroidal dipole. This can be achieved by inverting the time or the spatial coordinates. Thus, the polar toroidal dipole moment is antisymmetric under both space and time inversion.

The charge configuration in Fig. 2.1d that shall illustrate the axial toroidal dipole moment $\vec{g}$ will be discussed in the next section.

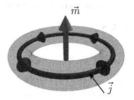

(a) Electric dipole moment $\vec{p}$ between a negative and a positive charge.

(b) Circular current $\vec{j}$ inducing a magnetic dipole moment $\vec{m}$.

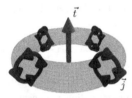

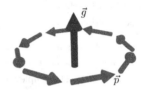

(c) Circular currents $\vec{j}$ on a torus inducing polar toroidal dipole moment $\vec{t}$.

(d) Circular arranged electric dipoles $\vec{p}$ inducing axial toroidal dipole moment $\vec{g}$.

Figure 2.1: The four principle charge-current configurations generating dipoles. After [28], [27] and [16].

2.3 Necessity for Three Multipole Families

In this section, we want to motivate why there is a necessity for three multipole families. Historically, for decades only two families were considered, namely the electric moments and the magnetic moments. This seemed sufficient, as the electric multipole moments were attributed to static electric charges, whereas the magnetic multipole moments were explained through moving electric charges. If we want to discuss what is missing in this picture, it is most instructive to discuss the symmetry properties of the electric and magnetic multipole families first and then to generalize this.

We will now consider the two discrete symmetries of parity and time inversion on the base of equations. As pointed out in the last section, the electric dipole moment behaves under spatial inversion like a polar vector, thus it holds [9, p. 271]

$$\mathcal{P}\vec{p} = -\vec{p}.$$

Here we used the parity operator $\mathcal{P}$, which is defined as such that it inverts spatial coordinates,

$$\mathcal{P}\vec{r} := -\vec{r}. \tag{2.1}$$

The magnetic dipole moment behaves under spatial inversion like a pseudovector, meaning that it remains invariant,

$$\mathcal{P}\vec{m} = \vec{m}.$$

Now we turn to the time inversion. The time reversal operator $\mathcal{T}$ is defined through

$$\mathcal{T}t := -t, \tag{2.2}$$

meaning that it changes the sign of the time variables. The electric dipole is the result of separated charges, and it holds [9, p. 271]

$$\mathcal{T}\vec{p} = \vec{p}.$$

Electric currents, as time derivatives of charges, change sign under time inversion [9, p. 271]. Hence, for the magnetic dipole moment caused by currents, it holds

$$\mathcal{T}\vec{m} = -\vec{m},$$

as it was made obvious in the last section.

Summarizing, $\vec{p}$ is even under time reversal, but odd under spatial inversion. $\vec{m}$ is odd under time reversal, but even under spatial inversion. The limitation to electric and magnetic moments thus misses a dipole moment which is odd under both time and spatial inversion, and another dipole moment, which is even under both transformations. We will illustrate this observation in the following table [16]:

Table 2.1: Behavior of the four dipoles under space and time inversion.

$\mathcal{T}$ \ $\mathcal{P}$	$+$	$-$
$+$	$\vec{g}$	$\vec{p}$
$-$	$\vec{m}$	$\vec{t}$

The "+" indicates even and the "−" odd behavior under the corresponding transformation. We introduced the dipole moments $\vec{g}$ and $\vec{t}$; we will call $\vec{g}$ the "axial toroidal dipole moment", and its multipole family "axial toroidal moments". The $\vec{t}$ is the "(polar) toroidal dipole moment", but in this thesis, we will from now on always refer to it only as "toroidal dipole moment". It is the lowest order of the (polar) toroidal multipole moment family. Toroidal moments interact only with time-dependent external electromagnetic fields [6], whereas electric and magnetic moments interact also with static external fields.

Following this table, one would expect that four multipole families are present in electrodynamic multipole expansions. An example of a charge distribution [16] generating an axial toroidal dipole moment is pictured in Fig. 2.1d. However, it has been argued [16, 17] that such a charge distribution is not stable in Maxwell-Lorentz electrodynamics. Therefore, axial toroidal multipole moments usually are considered as not possible in microscopic charge-current systems. For continuous media, the situation is different. There, axial toroidal moments can be generated e.g. by spin currents [30]. Also, such moments are suggested to appear in microscopic crystals with a specfic lattice structure [24]. Since we are discussing metamaterials rather than continuous media in this thesis, we will only consider (polar) toroidal moments in the following.

The previous reasoning was made for dipole moments only. One has to be careful when going to higher orders than the dipolar order. For example, the relation when applying the parity transformation to an arbitrary n-th order electric multipole moment tensor $\hat{P}^{(n)}$ is

$$\mathcal{P}\hat{P}^{(n)} = (-1)^n \hat{P}^{(n)}.$$

For arbitrary n-th poles, we have the following transformation rules [31, p. 227 and p. 257]:

Table 2.2: Behavior of the four general n-th pole moments under space and time inversion.

$\mathcal{T}$ \ $\mathcal{P}$	$(-1)^{n+1}$	$(-1)^n$
$+1$	$\hat{G}^{(n)}$	$\hat{P}^{(n)}$
-1	$\hat{M}^{(n)}$	$\hat{T}^{(n)}$

From the table it follows that n-th poles of the same order of electric and toroidal kind give the same fields that are not distinguishable with spatial manipulations of the experiment, since they share the same parity. But with respect to time it should be in principle possible to decide whether a charge-current distribution is characterized by an electric or toroidal n-th pole moment. However, no experiment has been carried out so far which provides such distinction.

2.4 Distinction between Toroidal Moment and Anapole

In literature, there is sometimes a confusion between the toroidal moment and the anapole [32]. The anapole is important in the theory of weak interaction where it was first postulated [14]. However, since we are interested in metamaterials rather than in particle physics, we focus here on the anapole description in classical electrodynamics. An anapole is in this context a charge-current distribution that neither radiates nor interacts with external fields [6].

An anapole can be realized in classical electrodynamics by a suitable combination of an electric and a toroidal dipole. Compared to the field of an electric dipole moment, the field of a toroidal dipole moment is scaled with the wave number k and has a phase-shift of $\pi/2$ relative to the field of the electric dipole [26]. Thus, by designing a charge-current configuration where the electric and toroidal dipoles point in the same direction and where the toroidal dipole exceeds the electric by a factor of k, the two fields annihilate each other exactly. In this configuration, however, the vector potential is non-zero, enabling Aharonov-Bohm like effects [19].

Because it does not radiate, the anapole is not a "moment" like the other multipole moments. This is the origin of the name "anapole" (from Greek 'ana', 'without', thus meaning 'without poles'). Nevertheless, the anapole is very often improperly denoted as "anapole moment" (e.g. in [32]). Because ideal non-radiating charge-current configurations are not possible [4], describing a structure as "non-radiating" refers only to the first orders of multipole moments. This is the reason why the anapole is usually only related with the toroidal dipole, not with higher toroidal moments.

3 Basic Equations and Notations

This chapter is dedicated to introduce the most important equations needed for the main part of the thesis. One of the basic tasks of electrodynamics is to find the electric and magnetic fields for given charge and current distributions. For this sake, Maxwell's equations are used, a system of partial inhomogeneous differential equations. To simplify the solution of these equations, the potential formalism is usually applied, including the scalar potential, the vector potential, and the Debye potentials. Furthermore, we will point out normalizations, symbols as well as sign and unit conventions used in this thesis.

3.1 Maxwell's Equations

Maxwell's equations are solved for given initial or boundary conditions. In differential form they are explicitly given as [9, p. 238]

$$\vec{\nabla} \cdot \vec{E}(\vec{r}, t) = \frac{1}{\varepsilon_0} \rho(\vec{r}, t) \,, \qquad (3.1) \qquad \vec{\nabla} \times \vec{E}(\vec{r}, t) = -\frac{\partial}{\partial t} \vec{B}(\vec{r}, t) \,, \qquad (3.3)$$

$$\vec{\nabla} \cdot \vec{B}(\vec{r}, t) = 0 \,, \qquad (3.2) \qquad \vec{\nabla} \times \vec{B}(\vec{r}, t) = \mu_0 \vec{j}(\vec{r}, t) + \mu_0 \varepsilon_0 \frac{\partial}{\partial t} \vec{E}(\vec{r}, t) \,. \qquad (3.4)$$

Equation (3.3) is also called Faraday's law and Eq. (3.4) is Maxwell's version of Ampere's law. Eq. (3.1) is Gauss's law and Eq. (3.2) is Gauss's law for magnetism.

$\vec{E}(\vec{r}, t)$ indicates the electric field, $\vec{B}(\vec{r}, t)$ the magnetic field, $\vec{j}(\vec{r}, t)$ the electric current density and $\rho(\vec{r}, t)$ the electric charge density. All quantities are given in SI units and evaluated at point $\vec{r}$ and time t. ε_0 is the vacuum permittivity and μ_0 the vacuum permeability. They are related via $\varepsilon_0 \mu_0 = 1/c^2$, where c is the speed of light in vacuum. Note the difference in the meaning of density: The current density is given as current per area, whereas the charge density has the units of charge per volume.

The above set of differential equations has to be completed with the continuity equation [9, p. 238],

$$\dot{\rho}(\vec{r}, t) + \vec{\nabla} \cdot \vec{j}(\vec{r}, t) = 0 \,, \qquad (3.5)$$

which relates the charge density of the system to the current which is flowing out of the system. In our systems, (3.5) is always fulfilled, but in the general case, the right side of the equation can have a non-zero value due to e.g. external fields and sources. The dot on top of ρ is the short notation of the time derivative $\partial/\partial t$. We will mainly use the dot in this thesis.

3.2 Wave Equation and Helmholtz Equation

Maxwell's equations describe the electromagnetic field, but it is not directly obvious that they also include an equation for describing an electromagnetic wave. This task is accomplished when Faraday's law (3.3) and Maxwell's version of Ampere's law (3.4) are combined, which yields two wave equations for the propagation of the electric and the magnetic field, respectively [9, p. 246]:

$$\Delta \vec{E}(\vec{r},t) - \frac{1}{c^2}\frac{\partial^2}{\partial t^2}\vec{E}(\vec{r},t) = \mu_0 \frac{\partial}{\partial t}\vec{j}(\vec{r},t) + \frac{1}{\varepsilon_0}\vec{\nabla}\rho(\vec{r},t)\,, \qquad (3.6)$$

$$\Delta \vec{B}(\vec{r},t) - \frac{1}{c^2}\frac{\partial^2}{\partial t^2}\vec{B}(\vec{r},t) = -\mu_0 \vec{\nabla} \times \vec{j}(\vec{r},t)\,. \qquad (3.7)$$

Since every time-dependent function can be represented as a decomposition into Fourier components [9, p. 407], we will use, where useful, harmonic time dependence $e^{-i\omega t}$, by just considering one term of the representation

$$\vec{j}(\vec{r},t) = \frac{1}{\sqrt{2\pi}}\int e^{-i\omega t}\vec{\tilde{j}}(\vec{r},\omega)\,\mathrm{d}\omega\,. \qquad (3.8)$$

Of course, the same decomposition is made for the charge density and all quantities derived from this, e.g. electric and magnetic fields. The sign of the argument of the exponential function is arbitrary, we use the minus sign in this thesis. Regarding the prefactor we choose the convention of symmetric prefactors $1/\sqrt{2\pi}$ for direct and inverse transformation. To keep notation clear, we will omit the tilde from now on at quantities given in the frequency domain; by means of the arguments, it is clear in which domain the quantity is given. To indicate that ω is more a parameter than a variable when considering only one term in the Fourier series, we also use the notation $\vec{j}_\omega(\vec{r})$ instead of $\vec{j}(\vec{r},\omega)$.

Because of Maxwell's equations, the change of the electromagnetic field in space and time is not independent. Both parts are connected through the dispersion relation

$$k^2(\omega) = \frac{\omega^2}{c^2}\varepsilon(\omega)\mu(\omega)\,, \qquad (3.9)$$

where k is the absolute value of the wave vector and $\varepsilon(\omega)$ and $\mu(\omega)$ are the permittivity and permeability in the frequency domain. For our purposes, it will be sufficient to assume $\varepsilon(\omega) = \mu(\omega) = 1$, so that $k = \omega/c$. Following this, we assume that $\vec{B} = \mu_0\vec{H}$ throughout the thesis.

Using the Fourier decomposition of type (3.8) in Eqs. (3.6) and (3.7), we arrive at the vector Helmholtz equations [9, p. 243]

$$[\Delta + k^2]\vec{E}(\vec{r}, \omega) = -i\omega\mu_0\vec{j}(\vec{r}, \omega) + \frac{1}{\varepsilon_0}\vec{\nabla}\rho(\vec{r}, \omega) \,, \tag{3.10}$$

$$[\Delta + k^2]\vec{B}(\vec{r}, \omega) = -\mu_0\vec{\nabla} \times \vec{j}(\vec{r}, \omega) \,. \tag{3.11}$$

Like in the case of the current, Eq. (3.8), the fields $\vec{E}(\vec{r}, \omega)$ and $\vec{B}(\vec{r}, \omega)$ are different from the $\vec{E}(\vec{r}, t)$ and $\vec{B}(\vec{r}, t)$ in Eqs. (3.6) and (3.7), but to keep notation clear we will stick to the notation $\vec{E}$ and $\vec{B}$ throughout the thesis. We will also omit the arguments in many cases. From context, it should become clear if the fields are in the frequency or in the time domain.

The homogeneous Helmholtz equations for spatial regimes without sources are

$$[\Delta + k^2]\vec{E}(\vec{r}, \omega) = 0 \,, \tag{3.12}$$

$$[\Delta + k^2]\vec{B}(\vec{r}, \omega) = 0 \,. \tag{3.13}$$

We will use them in chapter 5 to derive the Debye potentials and vector functions into which the fields $\vec{E}$ and $\vec{B}$ will be decomposed to simplify calculations.

3.3 Potentials

To determine the fields $\vec{E}$ and $\vec{B}$ in a simpler way than solving Eqs. (3.6) and (3.7), usually potentials are introduced. Sometimes it is easier to deal with them than with the fields. For the calculations in this thesis, we will only use the so-called "Lorenz gauge", because it has some advantages to be outlined in the following.

We start with discussing the electric and vector potential usually used. Hereafter, several other potentials which are used in this thesis are introduced.

3.3.1 Scalar and Vector Potential

Because of $\vec{\nabla} \cdot B = 0$, $\vec{B}$ can be represented through the curl of a vector field [9, p. 180],

$$\vec{B} = \vec{\nabla} \times \vec{A} \,. \tag{3.14}$$

$\vec{A}$ is called the vector potential, sometimes improperly also denoted as "magnetic vector potential" [33, 34]. However, the latter name is only appropriate in the static case, where the vector potential only causes the magnetic field and has contributions only from the magnetic multipoles. In the dynamic case, $\vec{A}$ contains contributions from all three multipole families and it also contributes to the electric field. We will therefore refer to $\vec{A}$ only as "vector potential" throughout the thesis.

With (3.14), Faraday's law (3.3) becomes

$$\vec{\nabla} \times \left(\vec{E} + \frac{\partial \vec{A}}{\partial t} \right) = 0 \,.$$

A field which has a vanishing curl can be expressed through the gradient of a scalar function,

$$\vec{E} + \frac{\partial \vec{A}}{\partial t} = -\vec{\nabla}\varphi \,. \tag{3.15}$$

φ is the electric scalar potential. We included the minus sign on the right side for convention [9, p. 30]. Following the textbooks [35, p. 9], we will usually call φ just "scalar potential", although we introduce some other scalar potentials in the next section.

The introduction of the two potentials reduces the originally six unknown quantities (three components of the electric and magnetic field, respectively) to four quantities (three components of $\vec{A}$ plus one of the scalar quantity φ). However, from the knowledge of the current $\vec{j}$ and hence the vector potential $\vec{A}$, anything else can be calculated, even if one has no information about ρ and the scalar potential φ [36]. Keeping this in mind, in this thesis we will mainly discuss the vector potential. We will refer to the scalar potential only if we want to point out important differences to the calculations of the vector potential or to illustrate an aspect at a simple example.

From definition (3.14) it follows that the vector potential is only defined up to the gradient of an arbitrary scalar potential. Thus, the vector potential can be redefined without changing the magnetic fields,

$$\vec{A} \rightarrow \vec{A} + \vec{\nabla}\chi \,,$$

where χ is an arbitrary function of $\vec{r}$ and t. Doing only this substitution would however change the electric field. To avoid this, one has to make the corresponding substitution for φ,

$$\varphi \rightarrow \varphi - \frac{\partial \chi}{\partial t} \,.$$

So both $\vec{A}$ and φ have the freedom of gauge. This fact can be exploited to simplify equations using a gauge suitable for the considered problem. Plugging in the fields expressed by the potentials yields equations for $\vec{A}$ and φ:

$$\Delta\varphi + \frac{\partial}{\partial t}(\vec{\nabla} \cdot \vec{A}) = -\frac{1}{\varepsilon_0}\rho\,, \tag{3.16}$$

$$\Delta\vec{A} - \frac{1}{c^2}\frac{\partial^2 \vec{A}}{\partial t} - \vec{\nabla}\left(\vec{\nabla} \cdot \vec{A} + \frac{1}{c^2}\frac{\partial\varphi}{\partial t}\right) = -\mu_0\vec{j}\,. \tag{3.17}$$

These four differential equations are, like Maxwell's equations, coupled, but can be decoupled by choosing the Lorenz gauge in free space, which is given by

$$\vec{\nabla} \cdot \vec{A} + \frac{1}{c^2}\frac{\partial\varphi}{\partial t} = 0\,. \tag{3.18}$$

We added "in free space", because some authors define the general Lorenz gauge with an additional term arising in continuous media [11, p. 240]. The Lorenz gauge has the advantage that the structure of the general solution for both potentials is, disregarding the different sources and numeric prefactors, exactly the same. So practically, numerical or analytical approximations need to be done only for one of the potentials and can then be applied to the other by using the fact that a solution of the scalar Helmholtz equation is also a solution of the vector Helmholtz equation [9, p. 429]. Additionally, in the Lorenz gauge every potential depends only on one kind of sources: The scalar potential depends only on static charges, whereas the vector potential depends only on currents, or, more general, on charges which change with time. These aspects simplify the calculations. However, one has to pay the price that the potentials do not have a direct physical meaning and are, if anything, only indirectly measurable.

The Lorenz gauge leads to

$$\Delta\varphi - \frac{1}{c^2}\frac{\partial^2 \varphi}{\partial t^2} = -\frac{1}{\varepsilon_0}\rho\,, \tag{3.19}$$

$$\Delta\vec{A} - \frac{1}{c^2}\frac{\partial^2 \vec{A}}{\partial t^2} = -\mu_0\vec{j}\,. \tag{3.20}$$

Equations (3.19) and (3.20) are solved by the retarded potentials

$$\varphi(\vec{r}, t) = \frac{1}{4\pi\varepsilon_0}\int \frac{\rho(\vec{r}', t - \frac{|\vec{r}-\vec{r}'|}{c})}{|\vec{r} - \vec{r}'|}\,\mathrm{d}^3 r'\,, \tag{3.21}$$

$$\vec{A}(\vec{r}, t) = \frac{\mu_0}{4\pi}\int \frac{\vec{j}(\vec{r}', t - \frac{|\vec{r}-\vec{r}'|}{c})}{|\vec{r} - \vec{r}'|}\,\mathrm{d}^3 r'\,. \tag{3.22}$$

The retarded time ensures that the cause is always before the effect, so that causality is preserved. This implies here that the effect of a change of the sources results in a change in the fields, which is the later the larger the distance is from the source to the point where the potential shall be

Figure 3.1: Toroidal vector $\vec{T}$ and poloidal vector $\vec{S}$. After [40].

evaluated. The speed at which the source change is transmitted to the evaluation point is the speed of light c.

The Lorenz gauge is the gauge we will use throughout this thesis. Hence, Eqs. (3.21) and (3.22) are the starting points for the multipole expansions based on the potentials in chapter 4.

3.3.2 Debye Potentials

The Debye potentials allow the calculation of the electromagnetic fields from three scalar potentials. By using the fundamental decomposition of a differentiable vector field into a divergence-free and a curl-free part [37, p. 733], also called Helmholtz decomposition [38], we can write

$$\vec{F} = \vec{\nabla}\xi + \vec{\nabla} \times \vec{V}.$$

The first term represents the curl-free part of $\vec{F}$, also called the longitudinal part; the second term accounts for the divergence-free part of $\vec{F}$, also called the transverse part. In three dimensions, the four unknown components of $\vec{V}$ and ξ can be reduced to three components by expressing the transverse part with two scalars by using the toroidal-poloidal decomposition [39]

$$\vec{\nabla} \times \vec{V} = \vec{S} + \vec{T} = -i(\vec{r} \times \vec{\nabla})\psi - i\vec{\nabla} \times (\vec{r} \times \vec{\nabla})\zeta.$$

The vectors $\vec{S}$ and $\vec{T}$ are pictured in Fig. 3.1. The prefactor $-i$ was included per convention, so that we can write the terms with the orbital angular momentum operator

$$\mathcal{L} = -i(\vec{r} \times \vec{\nabla}) \tag{3.23}$$

in a shorter way:

$$-i(\vec{r} \times \vec{\nabla})\psi - i\vec{\nabla} \times (\vec{r} \times \vec{\nabla})\zeta \equiv \mathcal{L}\psi + \vec{\nabla} \times \mathcal{L}\zeta.$$

$\vec{S}$ is also called "poloidal vector", $\vec{T}$ is sometimes denoted as "toroidal vector" [41]. These notions should not be confused with the toroidal moments. Toroidal currents induce toroidal electric fields and poloidal magnetic fields, and vice versa. In this sense, the term "toroidal electric field" has some ambiguity in its meaning: Either it corresponds to a toroidal current distribution and thus to a magnetic moment; or it is the effect of a toroidal moment, which would imply a poloidal current in the system. For the "toroidal magnetic field", the analogous confusion can arise. Because of this ambiguity, we will not use the term "toroidal field" in this thesis.

All in all, we get for a differentiable vector field the decomposition

$$\vec{F} = \vec{\nabla}\xi + \mathcal{L}\psi + \vec{\nabla} \times \mathcal{L}\zeta \, . \tag{3.24}$$

Hence, $\vec{F}$ is expressed through one longitudinal and two linearly independent transverse parts. This decomposition of $\vec{F}$ is both complete and unambiguous [39]. The scalar functions ξ, ψ and ζ are the so-called Debye potentials. They are determined as solutions of the scalar Helmholtz equation [42, p. 84]

$$[\Delta + k^2]f(\vec{r}, \omega) = 0 \, , \tag{3.25}$$

where f is one of the Debye potentials. In the special case of a rectangular coordinate system, every component of $\vec{E}$ and $\vec{B}$ satisfies the Helmholtz equation (3.25) directly [42, p. 59]. Because the term "scalar potential" is reserved for the electric scalar potential, we will refer to the Debye scalar potentials always as "Debye potentials".

4 Multipole Expansion of the Potentials

The integral representations of the scalar and vector potential, Eqs. (3.21) and (3.22), turn out to be unpractical for concrete calculations. Numerically, they can be evaluated for arbitrary source distributions, but such computations do not provide the insights to discuss the physical peculiarities of a particular source distribution. The relevant physics can best be made obvious by expanding a source distribution in a sum of specific contributions. Each of these contributions shall have a clear physical meaning. In this regard, the multipole expansion is a means of abstraction and provides a language to discuss the properties of source distributions. Performing a multipole expansion is in essence the approximation of the potentials or the fields of a source that is characterized by a specific charge or current distribution on a sphere enclosing the entire source, see Fig. 4.1.

In this chapter, we discuss several possibilities to perform the multipole expansion on the level of the potentials. The frequently used Taylor expansion in Cartesian coordinates is summarized in the first section. We show that this approach yields tensorial expressions which need to be converted into the multipole moment tensors using cumbersome formulae. To avoid this intricate procedure, we then demonstrate an alternative Cartesian expansion, which is based on the desire that the physically appropriate multipole tensors are fully symmetric and traceless. Hereby, an algorithm is outlined which enables to calculate toroidal multipole moments for arbitrary orders. The last section deals with the multipole expansion in spherical coordinates in the momentum space, not in the ordinary real space. This calculation is a possibility to calculate the vector potential of arbitrary multipole moments, and furthermore, this approach will suggest that the toroidal moments are in general not independent of the electric multipole moments.

Generally, we will discuss the dynamic case. If needed, one can simply take the limit $\omega \to 0$ to describe the static case as well. It can be shown that taking this limit always provides the correct result, namely the time-independent fields. [43, p. 703 et seq.]

4.1 Multipole Expansion in Cartesian Coordinates

This section shows briefly how the multipole expansion is usually done in textbooks. From this multipole expansion electric and magnetic moments will emerge, but no toroidal moments. We will show that this approach to the multipole expansion has some deficiencies. Particularly, **the multipole moment tensors in this expansion are not traceless**. This is undesirable, because non-traceless tensors have no definite properties under rotations, and the **toroidal moments are hidden in such non-traceless tensors**. Furthermore, the contribution of a certain expansion order will not just cause multipoles in the same order. More specifically, the

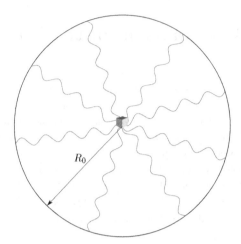

Figure 4.1: Two-dimensional scheme of the multipole expansion. A radiating object (the box in the middle of the circle) is enclosed into an imaginary sphere, on whose surface the potentials and fields (indicated by the wavy lines) are determined. For the Cartesian expansion, it is assumed that the radius R_0 of this sphere is much larger that the source itself.

dipole moments of electric and magnetic kind will be distributed into two expansion orders. Because this holds for all higher moments, too, the expansion gets quite messy, especially when toroidal moments and another quantity, the so-called "mean-square radii", are considered as well. We will show a possibility how the toroidal moments can be deduced from this expansion, but this will turn out to be quite cumbersome and not applicable for arbitrary high orders.

For the Taylor series we expand the potentials by using the parameter $|\vec{r}\,'|/|\vec{r}| \ll 1$. The idea is to truncate this generally infinity expansion after a few terms. The resulting approximation of the potential is the better the smaller the extent of the source, $|\vec{r}\,'|$, is in comparison to the distance $|\vec{r}|$ to the point where the potential shall be evaluated. The larger the source is, the more terms we have to take into account in our approximation. Furthermore, the Cartesian Taylor expansion is performed in the long-wavelength limit, meaning that the extent of the source is much smaller than the wavelength, $|\vec{r}\,'| \ll \lambda$ [9, p. 408], where λ is the wavelength of the incident electromagnetic wave and related to the wave number through $\lambda = 2\pi/k$. Whenever we will talk about "small sources", these assumptions are made.

4.1.1 Expanding the Scalar Potential

First, we want to expand the scalar potential, Eq. (3.21). We take advantage of the fact that the Taylor series of a product can be written as a Cauchy product of the two separate Taylor series,

$$\sum_{n=0}^{\infty} \frac{\vec{r}'^n}{n!} \vec{\nabla}_{\vec{r}'}^n \left. \frac{\rho\left(\vec{r}', t - |\vec{r}-\vec{r}'|/c\right)}{|\vec{r}-\vec{r}'|} \right|_{\vec{r}'=0}$$

$$= \sum_{n=0}^{\infty} \sum_{k=0}^{n} \frac{1}{n!} \vec{\nabla}_{\vec{r}'}^n \rho \left. \left(\vec{r}', t - \frac{|\vec{r}-\vec{r}'|}{c}\right) \right|_{\vec{r}'=0} \cdot \frac{1}{(n-k)!} \vec{\nabla}_{\vec{r}'}^{n-k} \left. \frac{1}{|\vec{r}-\vec{r}'|} \right|_{\vec{r}'=0} \vec{r}'^n. \quad (4.1)$$

Note that because we expand not directly in $\vec{r}'$, but in the small parameter $|\vec{r}'|/|\vec{r}|$, the Taylor series affects only the retarded time $t - |\vec{r}-\vec{r}'|/c$. The explicit spatial dependence $\vec{r}'$ of ρ just remains and is not expanded. The notation $\vec{\nabla}^n$ serves as an abbreviation, it holds

$$\vec{\nabla}^n = \vec{e}_1 \otimes \vec{e}_2 \otimes \ldots \otimes \vec{e}_n \frac{\partial}{\partial r_1} \frac{\partial}{\partial r_2} \cdots \frac{\partial}{\partial r_n} \quad (4.2)$$

for coordinates r_i, where $r_i = x, y, z$ and $\otimes$ denotes the matrix product of two vectors. So the result of this n-fold derivative is a tensor of rank n. The index of $\vec{\nabla}$ indicates the variable on which it is applied; if no index is given, it acts on $\vec{r}$.

For order n, the second factor in Eq. (4.1) can be simplified to

$$\vec{\nabla}_{\vec{r}'}^n \left. \frac{1}{|\vec{r}-\vec{r}'|} \right|_{\vec{r}'=0} = (-\vec{\nabla})_{\vec{u}}^n \left. \frac{1}{|\vec{u}|} \right|_{\vec{u}=\vec{r}} = (-\vec{\nabla})^n \frac{1}{r}. \quad (4.3)$$

The expansion of the charge density ρ yields

$$\sum_{n=0}^{\infty} \frac{\vec{r}'^n}{n!} \vec{\nabla}_{\vec{r}'}^n \rho \left. \left(\vec{r}', t - \frac{|\vec{r}-\vec{r}'|}{c}\right) \right|_{\vec{r}'=0} = \rho(\vec{r}', \tau) + \frac{r_i}{cr} \dot{\rho}(\vec{r}', \tau)$$

$$+ \left[\dot{\rho}(\vec{r}', \tau) \frac{r_i r_j - r^2 \delta_{ij}}{2cr^3} + \ddot{\rho}(\vec{r}', \tau) \frac{r_i r_j}{2c^2 r^2} \right] + \ldots . \quad (4.4)$$

For brevity, we introduced the retarded time $\tau = t - \frac{r}{c}$. The dot means derivative with respect to τ. Using Eqs. (4.1), (4.3) and (4.4), the Taylor expansion for the scalar potential, truncated after $n = 2$, is [10, p. 15 et seq.]

$$\varphi(\vec{r}, t) = \frac{1}{4\pi\varepsilon_0} \left\{ \frac{1}{r} \int \rho(\vec{r}', \tau) \, \mathrm{d}^3 r' + \frac{1}{r^3} \sum_i r_i \int r_i' \left(\rho(\vec{r}', \tau) + \frac{r}{c} \dot{\rho}(\vec{r}', \tau) \right) \mathrm{d}^3 r' \right.$$

$$+ \frac{1}{2} \sum_{ij} \int \left(\frac{3r_i r_j - r^2 \delta_{ij}}{r^5} \left(\rho(\vec{r}', \tau) + \frac{r}{c} \dot{\rho}(\vec{r}', \tau) \right) + \frac{r_i r_j}{c^2 r^3} \ddot{\rho}(\vec{r}', \tau) \right) r_i' r_j' \, \mathrm{d}^3 r'$$

$$\left. + \ldots \right\} . \quad (4.5)$$

This expansion looks rather messy, but by introducing some definitions for multipole moments, the notation can be clarified. We define now the following well-known tensors for the lowest orders by their components:

Electric monopole moment

$$q(\tau) = \int \rho(\vec{r}', \tau)\, \mathrm{d}^3 r'\,, \tag{4.6}$$

electric dipole moment

$$p_i(\tau) = \int r_i' \rho(\vec{r}', \tau)\, \mathrm{d}^3 r'\,, \tag{4.7}$$

primitive electric quadrupole moment

$$\tilde{Q}_{ij}^{(e)}(\tau) = \int r_i' r_j' \rho(\vec{r}', \tau)\, \mathrm{d}^3 r'\,. \tag{4.8}$$

The components of the electric multipole moment tensor of order n may be represented as [23]

$$\tilde{P}_{i_1 \ldots i_n}^{(n)}(\tau) = \int r_{i_1}' r_{i_2}' \ldots r_{i_n}' \rho(\vec{r}', \tau)\, \mathrm{d}^3 r' = \int \rho(\vec{r}', \tau) \prod_{m=1}^{n} r_{i_m}'\, \mathrm{d}^3 r'\,. \tag{4.9}$$

This form is sometimes called "primitive" moment [10, p. 4]. From the second order on, we note these moments with a tilde because it is not the suitable form to represent the electric multipole moments. However, in some textbooks the multipole moment tensors are defined in this way [10, p. 2 et seq.] [11, p. 18]. We will come back to this issue shortly. The monopole and dipole moments are the same for the different definitions, so the tilde is omitted.

Plugging the newly defined tensors into the expansion (4.5), we obtain the clearer terms [23]

$$
\begin{aligned}
\varphi(\vec{r}, t) &= \frac{1}{4\pi\varepsilon_0} \left\{ \frac{1}{r} q(\tau) + \frac{1}{r^3} \left(\vec{r} \cdot \vec{p}(\tau) + \frac{r}{c} \vec{r} \cdot \dot{\vec{p}}(\tau) \right) \right. \\
&\quad \left. + \frac{3 r_i r_j - r^2 \delta_{ij}}{2 r^5} \left(\tilde{Q}_{ij}^{(e)}(\tau) + \frac{r}{c} \dot{\tilde{Q}}_{ij}^{(e)}(\tau) \right) + \frac{r_i r_j}{2 c^2 r^3} \ddot{\tilde{Q}}_{ij}^{(e)}(\tau) + \ldots \right\} \\
&= \frac{1}{4\pi\varepsilon_0} \sum_{n}^{\infty} \frac{(-1)^n}{n!} \vec{\nabla}^n \left[\frac{1}{r} \hat{\tilde{P}}^{(n)}(\tau) \right]\,.
\end{aligned}
\tag{4.10}
$$

$\vec{\nabla}$ acts also on $\hat{\tilde{P}}^{(n)}(\tau)$ since the retarded time τ contains a spatial dependence. For the static case, meaning that all derivatives with respect to τ are zero, this equation simplifies to

$$
\begin{aligned}
\varphi(\vec{r}) &= \frac{1}{4\pi\varepsilon_0} \left(\frac{q}{r} + \frac{\vec{p} \cdot \vec{r}}{r^3} + \frac{3}{2 r^5} \vec{r} \cdot \hat{\tilde{Q}}^{(e)} \vec{r} - \frac{1}{2 r^3} \operatorname{Tr} \hat{\tilde{Q}}^{(e)} + \ldots \right) \\
&= \frac{1}{4\pi\varepsilon_0} \sum_{n}^{\infty} \frac{(-1)^n}{n!} \hat{\tilde{P}}^{(n)} \vec{\nabla}^n \frac{1}{r}\,.
\end{aligned}
\tag{4.11}
$$

4.1.2 Expanding the Vector Potential

Equivalent to expanding the scalar potential, this can be done with the vector potential (3.22) as well. Truncation after the second order yields in full analogy

$$\vec{A}(\vec{r}, t) = \vec{A}^{(0)}(\vec{r}, t) + \vec{A}^{(1)}(\vec{r}, t) + \vec{A}^{(2)}(\vec{r}, t) + \dots \tag{4.12}$$

with

$$\vec{A}^{(0)}(\vec{r}, t) = \frac{\mu_0}{4\pi} \frac{1}{r} \int \vec{j}(\vec{r}', \tau)\, \mathrm{d}^3 r', \tag{4.13}$$

$$\vec{A}^{(1)}(\vec{r}, t) = \frac{\mu_0}{4\pi} \frac{1}{r^3} \sum_i r_i \int r_i' \left(\vec{j}(\vec{r}', \tau) + \frac{r}{c} \dot{\vec{j}}(\vec{r}', \tau) \right) \mathrm{d}^3 r', \tag{4.14}$$

$$\vec{A}^{(2)}(\vec{r}, t) = \frac{\mu_0}{4\pi} \frac{1}{2} \sum_{ij} \int \left(\frac{3 r_i r_j - r^2 \delta_{ij}}{r^5} \left(\vec{j}(\vec{r}', \tau) + \frac{r}{c} \dot{\vec{j}}(\vec{r}', \tau) \right) + \frac{r_i r_j}{c^2 r^3} \ddot{\vec{j}}(\vec{r}', \tau) \right) r_i' r_j'\, \mathrm{d}^3 r'. \tag{4.15}$$

Now the question arises how we should define the multipole moment tensors. Of course, one could define the tensors directly in the order of their appearance in the expansion, similar to the definitions in the scalar potential. However, usually the terms are first manipulated using the continuity equation (3.5) and Gauss's theorem [9, p. 410]. Only after such manipulation the magnetic and electric multipole moment tensors are completely independent in every order. Then, from the multipole expansion of the vector potential in a specific order, an explicit contribution can be seen from the electric and magnetic multipole moment. For the zeroth order term (4.13), this procedure yields

$$\begin{aligned}
\vec{A}^{(0)}(\vec{r}, t) &= \frac{\mu_0}{4\pi} \frac{1}{r} \int \vec{j}(\vec{r}', \tau)\, \mathrm{d}^3 r' = -\frac{\mu_0}{4\pi} \frac{1}{r} \int \vec{r}' \left(\vec{\nabla}_{\vec{r}'} \cdot \vec{j}(\vec{r}', \tau) \right) \mathrm{d}^3 r' \\
&= \frac{\mu_0}{4\pi} \frac{1}{r} \int \vec{r}' \dot{\rho}(\vec{r}', \tau)\, \mathrm{d}^3 r' = \frac{\mu_0}{4\pi} \frac{1}{r} \dot{\vec{p}}(\tau),
\end{aligned} \tag{4.16}$$

where in the first step, we integrated by parts. So we see that electric multipole moments contribute to the expansion of the vector potential. The lowest order term is the electric dipole moment. Because no magnetic monopoles have been observed, a magnetic charge or current is not contained in Maxwell's equations and thus no monopole term is present in the vector potential. We omit the derivations for the first and second order terms here because we later do this for the general case of arbitrary order. Our definitions for the magnetic dipole and quadrupole moment are then:

Magnetic dipole moment

$$m_i(\tau) = \frac{1}{2} \int \left(\vec{r}' \times \vec{j}(\vec{r}, \tau) \right)_i \mathrm{d}^3 r', \tag{4.17}$$

primitive magnetic quadrupole moment

$$\tilde{Q}_{ij}^{(m)}(\tau) = \frac{2}{3} \int r_i' (\vec{r}' \times \vec{j}(\vec{r}',\tau))_j \, \mathrm{d}^3 r' , \tag{4.18}$$

and for arbitrary order n the primitive magnetic multipole moment tensor reads [23]

$$\tilde{M}_{i_1 \ldots i_n}^{(n)}(\tau) = \frac{n}{n+1} \int \left(\vec{r}' \times \vec{j}(\vec{r}',\tau) \right)_{i_n} \prod_{m=1}^{n-1} r_{i_m}' \, \mathrm{d}^3 r' . \tag{4.19}$$

Like the electric multipole moment tensor $\hat{\vec{P}}^{(n)}$, the magnetic multipole moment tensor $\hat{\vec{M}}^{(n)}$ is not traceless. An important difference is that $\hat{\vec{M}}^{(n)}$ is not fully symmetric in general when exchanging two arbitrary indices, because e.g.

$$\tilde{M}_{i_1 i_2 i_3}^{(3)} = \frac{3}{4} \int \left(\vec{r}' \times \vec{j}(\vec{r}',\tau) \right)_{i_1} r_{i_2}' r_{i_3}' \, \mathrm{d}^3 r' \neq \frac{3}{4} \int \left(\vec{r}' \times \vec{j}(\vec{r}',\tau) \right)_{i_2} r_{i_1}' r_{i_3}' \, \mathrm{d}^3 r' = \tilde{M}_{i_2 i_1 i_3}^{(3)} . \tag{4.20}$$

Contrary, $\hat{\vec{P}}^{(n)}$ is, as can be seen directly from Eq. (4.9), fully symmetric. As before, the tilde indicates that the definition of the primitive electric multipole moments is not the best choice to define the magnetic moments. The magnetic dipole moment is, like in the electric case, the same for all definitions, so the tilde can be omitted.

With the above definitions of primitive moments, the terms in the expansion (4.12) of the vector potential will take the following form [10, p. 17]:

$$\vec{A}^{(0)}(\vec{r},t) = \frac{\mu_0}{4\pi} \frac{1}{r} \dot{\vec{p}} , \tag{4.21}$$

$$\vec{A}^{(1)}(\vec{r},t) = \frac{\mu_0}{4\pi} \frac{1}{r^3} \left[\frac{1}{2} \dot{Q}_{ij}^{(e)} r_j - \epsilon_{ijk} r_j m_k + \frac{r}{2c} \ddot{Q}_{ij}^{(e)} r_j - \frac{r}{c} \epsilon_{ijk} r_j \dot{m}_k \right] \vec{e}_i , \tag{4.22}$$

$$\vec{A}^{(2)}(\vec{r},t) = \frac{\mu_0}{4\pi} \left\{ \frac{3 r_i r_k - r^2 \delta_{jk}}{2r^5} \left[\left(\frac{1}{3} \dot{O}_{ijk}^{(e)} - \epsilon_{ijl} \tilde{Q}_{lk}^{(m)} \right) + \frac{r}{c} \left(\frac{1}{3} \ddot{O}_{ijk}^{(e)} - \epsilon_{ijl} \dot{\tilde{Q}}_{lk}^{(m)} \right) \right] \right.$$
$$\left. + \frac{r_j r_k}{2c^2 r^3} \left(\frac{1}{3} \dddot{O}_{ijk}^{(e)} - \epsilon_{ijl} \ddot{\tilde{Q}}_{lk}^{(m)} \right) \right\} \vec{e}_i . \tag{4.23}$$

It is possible to continue Eq. (4.12) to arbitrary high orders, but there only primitive moments will appear and no toroidal moments at all, as we will show now. In fully analogy to the expansion of the scalar potential, Eq. (4.11), we are allowed to write for the vector potential [23]

$$\vec{A}(\vec{r},t) = \frac{\mu_0}{4\pi} \sum_{n=0}^{\infty} \frac{(-1)^n}{n!} \left[\prod_{m=2}^{n+1} \partial_{i_m} \right] \left[\frac{1}{r} \int \prod_{m=2}^{n+1} r_{i_m}' \, j_{i_1} \left(\vec{r}', \tau \right) \mathrm{d}^3 r' \right] \vec{e}_{i_1} . \tag{4.24}$$

For arbitrary n the following identity holds to express the last term:

$$\int \nabla_i \left[\prod_{m=1}^{n} r_{i_m} j_i \right] \mathrm{d}^3 r = \int \left[n j_{i_1} \prod_{m=2}^{n} r_{i_m} - \dot{\rho} \prod_{m=1}^{n} r_{i_m} + \sum_{k=2}^{n} \epsilon_{i_1 i_k l} (\vec{r} \times \vec{j})_l \prod_{\substack{m=2 \\ m \neq k}}^{n} r_{i_m} \right] \mathrm{d}^3 r . \quad (4.25)$$

The proof of this identity is lengthy and can be found in the appendix. Using Gauss's theorem [10, p. 211], it follows that the overall expression vanishes.

$$\int \nabla_i \left[\prod_{m=1}^{n} r_{i_m} j_i \right] \mathrm{d}^3 r = \int \prod_{m=1}^{n} r_{i_m} \vec{j} \cdot \mathrm{d}\vec{S} = 0 . \quad (4.26)$$

Hence, we can solve Eq. (4.25) for the first term on the right side:

$$\int j_{i_1} \prod_{m=2}^{n} r_{i_m} \, \mathrm{d}^3 r = \int \left[\frac{1}{n} \dot{\rho} \prod_{m=1}^{n} r_{i_m} - \frac{1}{n} \sum_{k=2}^{n} \epsilon_{i_1 i_k l} (\vec{r} \times \vec{j})_l \prod_{\substack{m=2 \\ m \neq k}}^{n} r_{i_m} \right] \mathrm{d}^3 r .$$

We now plug in this identity for $n \to n+1$ in Eq. (4.24) and simplify. This yields

$$\begin{aligned}
\vec{A}(\vec{r}, t) &= \frac{\mu_0}{4\pi} \sum_{n=0}^{\infty} \frac{(-1)^n}{n!} \left[\prod_{m=2}^{n+1} \partial_{i_m} \right] \left[\frac{1}{r} \int \prod_{m=2}^{n+1} r'_{i_m} j_{i_1}(\vec{r}', \tau) \, \mathrm{d}^3 r' \right] \vec{e}_{i_1} \\
&= \frac{\mu_0}{4\pi} \sum_{n=0}^{\infty} \frac{(-1)^n}{n!} \vec{\nabla}^n \left[\frac{1}{r} \frac{1}{n+1} \int \prod_{m=1}^{n+1} r'_{i_m} \dot{\rho}(\vec{r}', \tau) \, \mathrm{d}^3 r' \right. \\
&\qquad \left. - \frac{1}{n+1} \sum_{k=2}^{n+1} \epsilon_{i_1 i_k l} \int \left(\vec{r}' \times \vec{j}(\vec{r}', \tau) \right)_l \prod_{\substack{m=2 \\ m \neq k}}^{n+1} r'_{i_m} \, \mathrm{d}^3 r' \right] \vec{e}_{i_1} \\
&= \frac{\mu_0}{4\pi} \sum_{n=0}^{\infty} \frac{(-1)^n}{(n+1)!} \vec{\nabla}^n \left[\frac{1}{r} \dot{\hat{P}}^{(n+1)}(\tau) \right] \\
&\qquad - \frac{\mu_0}{4\pi} \sum_{n=1}^{\infty} \frac{(-1)^n}{n!} \vec{\nabla}^n \frac{1}{r} \left[\frac{1}{n+1} \sum_{k=2}^{n+1} \epsilon_{i_1 i_k l} \int \left(\vec{r}' \times \vec{j}(\vec{r}', \tau) \right)_l \prod_{\substack{m=2 \\ m \neq k}}^{n+1} r'_{i_m} \, \mathrm{d}^3 r' \right] \vec{e}_{i_1} \\
&= \frac{\mu_0}{4\pi} \sum_{n=1}^{\infty} \frac{(-1)^{n-1}}{n!} \vec{\nabla}^{n-1} \left[\frac{1}{r} \dot{\hat{P}}^{(n)}(\tau) \right] \\
&\qquad + \frac{\mu_0}{4\pi} \sum_{n=1}^{\infty} \frac{(-1)^{n-1}}{n!} \partial_{i_2} \cdots \partial_{i_{n+1}} \left[\frac{1}{r} \frac{1}{n} \sum_{k=2}^{n+1} \epsilon_{i_1 i_k l} \tilde{M}^{(n)}_{i_2 \ldots i_{k-1} i_{k+1} \ldots i_{n+1} l}(\tau) \right] \vec{e}_{i_1} \\
&= \frac{\mu_0}{4\pi} \sum_{n=1}^{\infty} \frac{(-1)^{n-1}}{n!} \vec{\nabla}^{n-1} \left[\frac{1}{r} \dot{\hat{P}}^{(n)}(\tau) \right] + \frac{\mu_0}{4\pi} \vec{\nabla} \times \sum_{n=1}^{\infty} \frac{(-1)^{n-1}}{n!} \vec{\nabla}^{n-1} \left[\frac{1}{r} \hat{M}^{(n)}(\tau) \right] \\
&= \frac{\mu_0}{4\pi} \sum_{n=1}^{\infty} \frac{(-1)^{n-1}}{n!} \vec{\nabla}^{n-1} \left[\frac{1}{r} \dot{\hat{P}}^{(n)}(\tau) + \vec{\nabla} \times \frac{1}{r} \hat{M}^{(n)}(\tau) \right] . \quad (4.27)
\end{aligned}$$

Thus, we recovered the primitive multipole moment tensors of order n, (4.9) and (4.19), as parts of an exact representation of the vector potential.

The representation (4.27) has a few pleasant properties: First, it shows that the vector potential in every order n can be written as a sum of the primitive magnetic n-th pole and the time derivative of the primitive electric n-th pole moment. Second, this form makes it easy to calculate an arbitrary order n of $\vec{A}$ just by calculating the tensors $\hat{\vec{P}}^{(n)}$ and $\hat{\vec{M}}^{(n)}$. It is also possible to express the n-th gradient of $1/r$ in a closed form [44].

Equation (4.27) contains no toroidal moments. Hence, the goal is now to bring it into a form in which the correct electric and magnetic multipole moment tensors and also the toroidal moments appear.

4.1.3 Tensorial Decomposition of the Vector Potential

In this section, we discuss a possibility to derive the toroidal dipole moment. This method is only applicable up to the second order term of the expansion (4.12) of the vector potential. For arbitrary high orders, we refer to the next section.

We exploit the fact that the physical appropriate tensors are symmetric and traceless. The property of symmetry ensures definite properties under parity transformation. The demand for tracelessness is connected with definite properties under rotations. Because the multipole expansion is, as mentioned, the approximation of the potential on a spherical surface, the functions which are contained in this expansion can be characterized by their properties regarding the transformations in the rotation group SO(3). These transformations are spatial rotations and the properties under such rotations can be classified with the total angular momentum value j. All dipole moments shall transform as a vector (meaning $j = 1$), all quadrupole moments shall transform as a tensor of rank 2 (meaning $j = 2$), and so on.

However, the trace of a multipole moment tensor is a scalar with respect to rotations. This would mix the properties under rotations when rotating non-traceless multipole moment tensors that contain different values of j. Also, a state with total angular momentum $j = 0$ is unphysical and does not occur in the fields. Mixing of angular momentum properties would make rotations hard to handle in calculations. Therefore, **a multipole moment tensor should contain only contributions from one total angular momentum value** j. Tensors that contain only one total angular momentum value are denoted as "irreducible" tensors. Tensors which have mixed properties under SO(3) (i.e. more than one j) are denoted as "reducible" tensors [45]. Irreducible tensors are traceless. **Our goal is therefore to decompose the reducible primitive multipole moment tensors appearing in the expansion of the vector potential (4.27) into irreducible tensors.**

Traceless tensors have less independent entries than non-traceless tensors. This means that the primitive multipole moment tensors are overdetermined regarding the complete description of a given potential. We show this as follows: The electric multipole moment tensor of order j has $N = 3^j$ entries, but because the tensor is fully symmetric, only

$$N_{\text{sym}}(j) = \frac{1}{2}(j+1)(j+2)$$

entries are independent [46]. We now impose tracelessness, meaning that it holds

$$\sum_k \sum_{i_\alpha, i_\beta} T^{(j)}_{i_1 i_2 i_3 \ldots i_j} \delta_{i_\alpha k} \delta_{i_\beta k} = 0$$

for an traceless tensor $\hat{T}$, an arbitrary choice of i_α and i_β out of the set of indices $i_1 \ldots i_j$. This reduces the number of independent entries further.

The number of equations which limit the degrees of freedom for the entries of a tensor can be calculated as follows: A tensor of rank two has exactly one trace, this yields one equation to restrict the degrees of freedom of the tensor's entries. For a tensor of rank three there are three possibilities to calculate the trace, and for a tensor of rank j, there are

$$N_{\text{t}}(j) = \sum_{k=1}^{j-1} k = \frac{1}{2}j(j-1)$$

traces. Thus, the number of independent entries of a symmetric and traceless rank-j-tensor is [47]

$$\begin{aligned}
N_{\text{sym-trl}}(j) &= N_{\text{sym}}(j) - N_{\text{t}}(j) \\
&= \frac{1}{2}(j+1)(j+2) - \frac{1}{2}j(j-1) \\
&= 2j+1 \,.
\end{aligned} \tag{4.28}$$

Thus, we expect $2j+1$ independent entries for a tensor of rank j. This is consistent with the number of values to which the projection of the orbital angular momentum onto the z-axis is restricted for a given angular momentum eigenvalue j. It is always possible to choose a traceless representation for a multipole moment tensor; for a given non-traceless tensor, the trace can simply be subtracted and added to a multipole moment tensor of lower rank [46]. Accordingly, it is necessary to start detracing a tensor from the highest order considered, and to subsequently detrace the tensors going from higher to lower orders [23].

To define the multipole moment tensors in a traceless form, we consider the first order term of the Taylor series of the vector potential, Eq. (4.14),

$$\vec{A}^{(1)}(\vec{r},t) = \frac{\mu_0}{4\pi} \frac{1}{r^3} \sum_i r_i \int r_i' \left(\vec{j}(\vec{r}',\tau) + \frac{r}{c}\dot{\vec{j}}(\vec{r}',\tau) \right) \mathrm{d}^3 r' \,.$$

To shorten the notation, we suppress the arguments of $\vec{j}$ and reduce the expression to the basic tensor which is contained in both terms,

$$T_{ij} = \int r'_i j_j \, \mathrm{d}^3 r' \,. \tag{4.29}$$

The symmetry properties of this tensor are unknown in general. The idea is now to express this tensor as a sum of tensors with known symmetry properties. Every tensor of rank two can be decomposed into a symmetric and an antisymmetric part [37, p. 287]. The symmetric part can further be decomposed into a symmetric traceless tensor and the trace. So we have the general decomposition [48]

$$\hat{T} = \hat{T}^{(a)} + \hat{T}^{(st)} + \hat{T}^{(t)} \,. \tag{4.30}$$

$\hat{T}^{(a)}$ is the antisymmetric part, $\hat{T}^{(st)}$ the symmetric traceless part and $\hat{T}^{(t)}$ contains the trace. $\hat{T}^{(st)}$ is also called "natural representation" of the tensor $\hat{T}$ [49]. The components of these tensors are defined according to

$$T_{ij}^{(a)} = \frac{1}{2}(T_{ij} - T_{ji}) \,, \tag{4.31}$$

$$T_{ij}^{(t)} = \delta_{ij} \operatorname{Tr} \hat{T} \tag{4.32}$$

and

$$T_{ij}^{(st)} = \frac{1}{2}(T_{ij} + T_{ji}) - \delta_{ij} \operatorname{Tr} \hat{T} \,. \tag{4.33}$$

These three tensors are the irreducible tensors constituting the second order of the expansion of the vector potential. Obviously, the fully symmetric traceless tensor behaves under exchange of two indices like

$$T_{ij}^{(st)} = T_{ji}^{(st)} \,,$$

the antisymmetric tensor like

$$T_{ij}^{(a)} = -T_{ji}^{(a)} \,.$$

These symmetry properties are related to the parity properties of the multipole moments discussed in section 2.2. The symmetric traceless part will yield the electric quadrupole tensor, which is, following Tab. 2.2, even under parity, whereas the antisymmetric part will yield the symmetric magnetic dipole moment times the fully antisymmetric Levi-Civita tensor.

To manipulate the tensor in Eq. (4.29), we use the continuity equation (3.5) to perform manipulations like in Eq. (4.16). The antisymmetric tensor is

$$
\begin{aligned}
T_{ij}^{(a)} &= \frac{1}{2} \int \left(r_i' j_j - r_j' j_i \right) \mathrm{d}^3 r' = \frac{1}{2} \epsilon_{nij} \int \epsilon_{nmp} r_m' j_p \, \mathrm{d}^3 r' \\
&= \frac{1}{2} \epsilon_{nij} \int (\vec{r}' \times \vec{j})_n \, \mathrm{d}^3 r' = -\epsilon_{jin} m_n \, .
\end{aligned}
\tag{4.34}
$$

$\vec{m}$ is the magnetic dipole moment defined in Eq. (4.17). The symmetric traceless part is

$$
\begin{aligned}
T_{ij}^{(st)} &= \frac{1}{2} \int \left(r_i' j_j + r_j' j_i - \frac{1}{3} r_m' j_m \delta_{ij} \right) \mathrm{d}^3 r' = \frac{1}{2} \int \left(r_j' r_i' \partial_m j_m - \frac{1}{3} r_n' r_n' \partial_m j_m \delta_{ij} \right) \mathrm{d}^3 r' \\
&= \frac{1}{2} \int \left(r_i' r_j' - \frac{1}{3} r'^2 \delta_{ij} \right) \dot{\rho} \, \mathrm{d}^3 r' = \frac{1}{2} \dot{Q}_{ij}^{(e)} \, ;
\end{aligned}
\tag{4.35}
$$

$$
Q_{ij}^{(e)}(\tau) = \int \left(r_i' r_j' - \frac{1}{3} r'^2 \delta_{ij} \right) \rho(\vec{r}', \tau) \, \mathrm{d}^3 r'
\tag{4.36}
$$

is the traceless electric quadrupole moment. This form, but with a different normalization, can be found in standard textbooks [9, p. 414]. We denote it without the tilde to distinguish it from the primitive quadrupole moment (4.8).

Finally, the trace is

$$
\begin{aligned}
T_{ij}^{(t)} &= \frac{1}{3} \delta_{ij} \int r_j' j_j \, \mathrm{d}^3 r' = \frac{1}{3} \delta_{ij} \int \frac{1}{2} r_j' r_j' \partial_m j_m \, \mathrm{d}^3 r' \\
&= \frac{1}{6} \delta_{ij} \int r'^2 \dot{\rho} \, \mathrm{d}^3 r' = \frac{1}{6} \dot{r}_q^{(2)} \delta_{ij} \, .
\end{aligned}
\tag{4.37}
$$

Here we defined the mean-square radius of the charge distribution,

$$
r_q^{(2)}(\tau) = \int (r')^2 \rho(\vec{r}', \tau) \, \mathrm{d}^3 r' \, .
\tag{4.38}
$$

In general, the charge mean-square radius of order n is given as [16]

$$
r_q^{(2n)} = \int (r')^{2n} \rho(\vec{r}', \tau) \, \mathrm{d}^3 r' \, .
\tag{4.39}
$$

Using this explicit decomposition of $\hat{T}$, we can write $\vec{A}^{(1)}(\vec{r}, t)$ as follows:

$$
\begin{aligned}
\vec{A}^{(1)}(\vec{r}, t) &= \frac{\mu_0}{4\pi} \frac{1}{r^3} \left[\frac{1}{2} \dot{Q}^{(e)}(\tau) \cdot \vec{r} + \frac{r}{2c} \ddot{Q}^{(e)}(\tau) \cdot \vec{r} + \vec{m}(\tau) \times \vec{r} + \frac{r}{c} \dot{\vec{m}}(\tau) \times \vec{r} \right. \\
&\quad \left. + \frac{1}{6} \left(\dot{r}_q^{(2)}(\tau) + \frac{r}{c} \ddot{r}_q^{(2)}(\tau) \right) \vec{r} \right] .
\end{aligned}
\tag{4.40}
$$

So we see that the mean-square radius $r_q^{(2)}(\tau)$ does not affect the magnetic field [10, p. 27] because $\vec{\nabla} \times f(r)\, \vec{r} = 0$ holds for every differentiable function $f(r)$ that only depends on the absolute value of $\vec{r}$. $r_q^{(2)}(\tau)$ does also not change the electric field, because using Eq. (3.15) both mean-square radii from the scalar potential (not shown here) and vector potential cancel each other. Mean-square radii of the charge distribution occur e.g. in a spherical capacitor. In such a system, all multipole moments are zero, but the charge mean-square radii have finite values [16].

Because normally, one is only interested in the physical observable fields, the electric mean-square radii (i.e. mean-square radii of the electric multipole moments) for exactly this reason are usually omitted in textbooks [9, p. 413 et seq.]. Hence, the electric mean-square radii can be viewed as a gauge artifact, which can be removed by a different choice of $\vec{A}$ and φ. Therefore, it does not matter in the first order of the vector potential if the primitive or the traceless multipole moment tensors are used. In higher orders, this is not true. There, **it makes a difference in the fields if one uses the primitive or the traceless multipole moment tensors** [10, p. 28]. We will discuss this in chapter 6. This difference is caused by the toroidal moments, but also by mean-square radii of the magnetic and toroidal multipole moments. In contrast to almost all electric mean-square radii (the order $n = 0$ of $r_q^{(2n)}$ corresponds to the normal radiating multipole moments), the magnetic and toroidal mean-square radii radiate and therefore have to be considered for the radiation field [22].

All in all, we found in the first order of the vector potential a contribution from the electric quadrupole moment and the magnetic dipole moment together with the first mean-square radius of the charge distribution. This is in some sense undesirable, it would be better if in the order n only n-th poles of all three multipole families appear. This is not the case because the Cartesian Taylor expansion relies on the parameter r'/r, and the magnetic multipoles are always one order higher with respect to this parameter than the electric multipoles. Taking the toroidal moments into account, they appear even one order higher, with the lowest moment, the toroidal dipole, appearing in the second order, together with the electric octupole and the magnetic quadrupole. To show this, we consider the second order term, Eq. (4.15), in the expansion of the vector potential,

$$A_i^{(2)}(\vec{r},t) = \frac{\mu_0}{4\pi}\frac{1}{2}\sum_{jk}\int \left(\frac{3r_j r_k - r^2 \delta_{jk}}{r^5}\left(j_i(\vec{r}',\tau) + \frac{r}{c}\dot{j}_i(\vec{r}',\tau)\right) + \frac{r_j r_k}{c^2 r^3}\ddot{j}_i(\vec{r}',\tau)\right) r_j' r_k'\, \mathrm{d}^3 r'.$$

(4.41)

In principle, the idea to decompose this tensor in several symmetric and antisymmetric parts is the same as before, but because of the higher rank, the formulae to calculate the individual constituent tensors are much lengthier [50]. We only give here the result of the decomposition, the calculation can be found in the appendix.

$$
\begin{aligned}
\vec{A}^{(2)}(\vec{r},t) = \frac{\mu_0}{4\pi}\frac{1}{2r^5}\Bigg[& \vec{r}^T \dot{O}^{(e)} \cdot \vec{r} + \frac{r}{c}\vec{r}^T \ddot{O}^{(e)} \cdot \vec{r} + \frac{r^2}{3c^2}\vec{r}^T \dddot{O}^{(e)} \cdot \vec{r} - 2\vec{r}\times(\hat{Q}^{(m)}\cdot\vec{r}) \\
& -\frac{2r}{c}\vec{r}\times(\dot{\hat{Q}}^{(m)}\cdot\vec{r}) - \frac{2r^2}{3c^2}\vec{r}\times(\ddot{\hat{Q}}^{(m)}\cdot\vec{r}) + 6\vec{r}\,(\vec{r}\cdot\vec{t}) - 2r^2\vec{t} + \frac{6r}{c}\vec{r}\,(\vec{r}\cdot\dot{\vec{t}}) - \frac{2r^3}{c}\dot{\vec{t}} \\
& +\frac{2r^2}{c^2}\vec{r}\,(\vec{r}\cdot\ddot{\vec{t}}) + \frac{9}{5}\vec{r}\,(\vec{r}\cdot\vec{r}_{\vec{p}}^{(2)}) + \frac{3r}{5c}\vec{r}\,(\vec{r}\cdot\dot{\vec{r}}_{\vec{p}}^{(2)}) + \frac{3r^2}{5c^2}\vec{r}\,(\vec{r}\cdot\ddot{\vec{r}}_{\vec{p}}^{(2)}) - \frac{r^2}{5}\vec{r}_{\vec{p}}^{(2)} - \frac{r^3}{5c}\dot{\vec{r}}_{\vec{p}}^{(2)} \Bigg]
\end{aligned}
$$

$$(4.42)$$

The following quantities have been defined:

Electric octupole moment

$$
O_{ijk}^{(e)}(\tau) = \int \left(r_i' r_j' r_k' - \frac{1}{5}r'^2(r_i'\delta_{jk} + r_j'\delta_{ik} + r_k'\delta_{ij}) \right) \rho(\vec{r}',\tau)\,\mathrm{d}^3 r', \tag{4.43}
$$

magnetic quadrupole moment

$$
Q_{ij}^{(m)}(\tau) = \frac{1}{3}\int \left[(\vec{j}(\vec{r}',\tau)\times\vec{r}')_i r_j' + (\vec{j}(\vec{r}',\tau)\times\vec{r}')_j r_i' \right]\mathrm{d}^3 r', \tag{4.44}
$$

toroidal dipole moment

$$
t_i(\tau) = \frac{1}{10}\int \left[r_i'(\vec{r}'\cdot\vec{j}(\vec{r}',\tau)) - 2(r')^2 j_i(\vec{r}',\tau) \right]\mathrm{d}^3 r', \tag{4.45}
$$

mean-square radius of the electric dipole moment

$$
\vec{r}_{\vec{p}}^{(2)}(\tau) = \int (r')^2\,\vec{r}'\rho(\vec{r}',\tau)\,\mathrm{d}^3 r'. \tag{4.46}
$$

Thus, we encounter in this decomposition contributions from the traceless electric octupole moment and the traceless symmetric magnetic quadrupole moment. Additionally, we find the toroidal dipole moment and another mean-square radius, this time the one belonging to the electric dipole moment. As mentioned before, this mean mean-square radius does also not contribute to the fields. The definition of the toroidal dipole moment $\vec{t}$ in this derivation is a bit artificial, as it was chosen as the simplest transverse vector and with the constraint that also the mean-square radius of the electric dipole has to occur in this order of the expansion [16].

The following table illustrates in which orders n the different multipole moments arise from a Taylor expansion with subsequent decomposition of the tensors:

The left column gives the type of the multipole moment (electric, magnetic, or toroidal). We see that to get the three dipoles $\vec{p}$, $\vec{m}$, and $\vec{t}$, we need to calculate three orders of the vector potential.

Table 4.1: Overview of the multipole moments emerging in different orders of $\vec{A}$

type \ n	0	1	2
el.	$\vec{p}$	$\hat{Q}^{(e)}$	$\hat{O}^{(e)}$
mag.	-	$\vec{m}$	$\hat{Q}^{(m)}$
tor.	-	-	$\vec{t}$

The demonstrated way to decompose the terms of the vector potential expansion is not feasible in practice for arbitrary high orders [9, p. 415]. Already for the third order, it is necessary to decompose a fourth rank tensor, and the formulae needed for this fill several pages [47]. For the fourth and all higher orders, no formulae for the general case have been derived. So we need to look for other possibilities to derive the toroidal moments and to express the vector potential with its fundamental symmetric constituents.

We can summarize that the Cartesian Taylor expansion has several inconveniences: We have the limitation that the sources are confined to a small spatial region compared to the distance where the potential shall be evaluated. Furthermore, the results are only valid in the long-wavelength limit and the emerging tensors have no definite properties under rotations and parity. Furthermore, the n-th poles do not appear in the same order n: The electric n-th pole always appears in the same order as the magnetic $(n-1)$-th pole, and the toroidal n-th pole appears in the same order as the electric $(n+2)$-th pole. Last but not least, a decomposition of the Cartesian vector potential in terms of irreducible tensors is not doable in practice for arbitrary high orders.

To resolve or completely avoid these problems, there are two possibilities: One of them uses the Cartesian coordinates and is based on the idea that physical useful multipole moments behave properly under parity and rotations and therefore, following the above discussion, have to be symmetric and traceless. Still, this comes with the restriction of spatially confined sources, but the problems of non-traceless tensors are solved. The other approach uses spherical coordinates and a decomposition of the current density in momentum space. With this, it is possible to calculate the multipole moments on the base of angular momentum eigenstates.

At first, we will discuss at first the algorithm which exploits the demand of symmetry and tracelessness of the multipole tensors. Hereafter, the expansion in spherical coordinates and momentum space is analyzed.

4.2 Cartesian Traceless Expansion including Toroidal Moments

The primitive moments of order n contain irreducible parts which do not transform as a tensor of rank n. In the last section, we showed for the first and second order term of the vector potential, how the toroidal moments can be defined by decomposing the vector potential. In this section we discuss an algorithm first proposed in [23] how the vector potential can be expressed with electric, magnetic and toroidal multipole moments in arbitrary orders.

4.2.1 From Primitive Moments to Traceless Moments

The essential idea which is outlined in this section is the following: We are going to take the two primitive multipole moment tensors $\hat{\vec{P}}$ and $\hat{\vec{M}}$ and manipulate them in a way that three new tensors emerge from the calculation: the traceless electric and magnetic multipole moment tensors as well as the toroidal multipole moment tensor. **These three new tensors will, unlike the two former primitive tensors, represent distinct physical symmetry properties of the multipole field.** More specifically, they will fulfill the properties outlined in section 2.2. This procedure was first proposed in 2002 [23] and refined in 2005 [51].

To construct the traceless multipole moments, we will use an operator which takes the trace out of a tensor with arbitrary rank n [52]. It is self-evident to call this operator "detracing operator" $\mathcal{D}$. There is one restriction to the tensors on which $\mathcal{D}$ can be applied: They have to be totally symmetric, i.e. symmetric in exchanging an arbitrary pair of indices. This applies for the electric multipole moment tensor (4.9), but not for the magnetic multipole moment tensor (4.19). Actually, $\hat{\vec{M}}$ contains contributions which behave under a parity transformation like electric multipole moments. We will remove these parts from $\hat{\vec{M}}$ by symmetrizing it. These parts with the same parity as the electric multipole moments will later be attributed to the toroidal multipole moments.

One important feature of this transformation is that it cannot be done for every order separately in the dynamic case. This is only possible in the static case, but then, no toroidal moments are present and the procedure to detrace the tensors is both trivial and unnecessary.

Furthermore, one has to decide up to which order N the traceless tensors, including the toroidal moments, shall be derived, and then consider all terms of the vector potential (4.27) until the order $N = n + 2$, where the sum index n enumerates the order of the Taylor expansion. This is due to the fact that the physical fields $\vec{E}$ and $\vec{B}$ up to the order n are required not to change when performing this transformation, meaning that the sums

$$\sum_{l \leq n} \vec{E}^{(l)} \quad \text{and} \quad \sum_{l \leq n} \vec{B}^{(l)}$$

remain invariant. The requirement not to alter these sums by detracing the tensors in a certain order leads to the necessity to add compensation terms in a different, lower order.

The calculation of the symmetric part of the primitive magnetic multipole moment tensor $\hat{\tilde{M}}$ is a generalization of Eq. (4.31) for the case of arbitrary tensor rank,

$$\tilde{M}_{i_1 \ldots i_n}^{(s)} = \tilde{M}_{i_1 \ldots i_n} - \frac{1}{n} \sum_{p=1}^{n-1} [\tilde{M}_{i_1 \ldots i_{p-1} i_{p+1} \ldots i_{n-1} i_p i_n} - \tilde{M}_{i_1 \ldots i_{p-1} i_{p+1} \ldots i_{n-1} i_n i_p}] . \qquad (4.47)$$

Using the identity [10, p. 211]

$$\epsilon_{ijk} \epsilon_{imn} = \delta_{jm} \delta_{kn} - \delta_{jn} \delta_{km} ,$$

we rewrite Eq. (4.47) as

$$\tilde{M}_{i_1 \ldots i_n}^{(s)} = \tilde{M}_{i_1 \ldots i_n} - \frac{1}{n} \sum_{p=1}^{n-1} \epsilon_{i_p i_n q} \tilde{N}_{i_1 \ldots i_{p-1} i_{p+1} \ldots i_{n-1} q} . \qquad (4.48)$$

For abbreviation, we introduced the tensor $\hat{\tilde{N}}$, defined by its components [23]

$$\begin{aligned} \tilde{N}_{i_1 \ldots i_n}(\tau) &= \sum_{ps} \epsilon_{i_n ps} \tilde{M}_{i_1 \ldots i_{n-1} ps}(\tau) \\ &= \sum_{ps} \frac{n+1}{n+2} \epsilon_{i_n ps} \int r'_{i_1} \ldots r'_{i_{n-1}} r'_p [\vec{r}' \times \vec{j}(\vec{r}', \tau)]_s \, \mathrm{d}^3 r' \\ &= \frac{n+1}{n+2} \int r'_{i_1} \ldots r'_{i_{n-1}} [\vec{r}' \times (\vec{r}' \times \vec{j}(\vec{r}', \tau))]_{i_n} \, \mathrm{d}^3 r' . \end{aligned} \qquad (4.49)$$

The tilde indicates that this tensor is not fully symmetric and, like the magnetic multipole moment tensor, needs to be symmetrized. $\hat{\tilde{N}}$ is that part of $\hat{\tilde{M}}$ which has the opposite parity when compared to the physical relevant magnetic moments described in chapter 2. For example,

$$\hat{N}^{(1)}(\tau) = \frac{2}{3} \vec{N}(\tau) = \int \vec{r}' \times [\vec{r}' \times \vec{j}(\vec{r}', \tau)] \, \mathrm{d}^3 r'$$

and thus

$$\mathcal{P} \vec{N}(\tau) = \frac{2}{3} \int (-\vec{r}') \times [-\vec{r}' \times (-\vec{j}(\vec{r}', \tau))] \, \mathrm{d}^3 r' = -\vec{N}(\tau) ,$$

and in general

$$\mathcal{P} \hat{N}^{(n)} = (-1)^n \hat{N}^{(n)} .$$

We used the notation without the tilde to indicate that $\hat{N}$ is the symmetrized version of $\hat{\tilde{N}}$ by making use of the symmetrizing relation (4.47). Contrary, it holds $\mathcal{P} \vec{m} = \vec{m}$ and in general $\mathcal{P} \hat{M}^{(n)} = (-1)^{n+1} \hat{M}^{(n)}$. This means that $\vec{N}$ (and in general all $\hat{N}^{(n)}$) has the same parity as the electric and toroidal multipole moments. Because we directly have, using the behavior of the electric current under time inversion,

$$\mathcal{T}\vec{N} = \frac{2}{3}\int \vec{r}' \times [\vec{r}' \times (-\vec{j}(\vec{r}',\tau))]\, \mathrm{d}^3 r' = -\vec{N}(\tau)\,,$$

and in general $\mathcal{T}\hat{N} = -\hat{N}$, the parity under time inversion is the same as for toroidal moments. We will see later that the toroidal dipole moment will be composed partially from $\vec{N}$. This vector was actually the term Van Bladel discussed in 1988 [18] without referring to toroidal multipole moments. For some considerations, even nowadays only such a term is denoted as toroidal dipole moment [53].

Now that we have the fully symmetric tensors $\hat{\hat{P}}$ and $\hat{\hat{M}}^{(s)}$, we can go on and detrace them. The detracing operator $\mathcal{D}^{(n)}$ is defined for a totally symmetric tensor $\hat{T}^{(n)}$ through

$$\mathcal{D}^{(n)}\hat{T} = \hat{T}^{(st)}\,. \tag{4.50}$$

$\mathcal{D}^{(n)}$ projects out of a general totally symmetric rank-n-tensor exactly this part which transforms as a tensor with total angular momentum $j = n$. Because it also holds $\mathcal{D}^{(n)}\mathcal{D}^{(n)} = \mathcal{D}^{(n)}$, $\mathcal{D}^{(n)}$ is a projector [52]. The entries of $\hat{T}^{(st)}$ are given by

$$T^{(st)}_{i_1\dots i_n} = T_{i_1\dots i_n} - \sum_{D(i)} \delta_{i_1 i_2}\Lambda[\hat{T}^{(n)}]_{i_3 i_4\dots i_n} \tag{4.51}$$

with the tensorial functional $\hat{\Lambda}$ that takes as argument a totally symmetric tensor of rank n and yields as result a tensor of rank $n-2$,

$$\Lambda[\hat{T}^{(n)}]_{i_3\dots i_n} = \sum_{m=1}^{\lfloor n/2 \rfloor} \frac{(-1)^{m-1}(2n-1-2m)!!)}{(2n-1)!!\, m} \sum_{D(i)} \delta_{i_3 i_4\dots i_{2m-1} i_{2m}} T^{n;m}_{i_{2m+1}\dots n}\,. \tag{4.52}$$

$\lfloor n/2 \rfloor$ is the integer value equal to or rounded down from $n/2$. $!!$ is the double factorial and defined for odd n as $n!! = n \cdot (n-2) \cdot (n-4) \cdot \dots \cdot 1$. The superscript n is the rank of the tensor which is to be detraced, and the superscript m is the number of contractions. For one contraction, two arbitrary indices are set equal and then those components are summed up. In this sense, a contraction is the general case of a trace. The result of one contraction is a tensor of rank $n-2$. $\delta_{i_3 i_4\dots i_{2m-1} i_{2m}}$ is a delta function with two of the indices that appear on the left side of the equation. These two indices are shuffled with in the sum $\sum_{D(i)}$. It is the sum of all permutations regarding the indices of a tensor,

$$\sum_{D(i)} T_{i_1\dots i_n} = \frac{1}{n}[T_{i_1\dots i_n} + T_{i_n\dots i_1} + T_{i_1 i_n\dots i_2} + \dots + T_{i_1\dots i_n i_{n-1}}]\,.$$

The tensor Λ contains the traces of $\hat{T}$ in a form that it transforms as a tensor of rank $n-2$. So for $n = 3$, Λ is as a vector. The parity of Λ regarding spatial and time inversions will be the same as the tensor from which the traces are calculated. Thus, when calculating e.g. the traces of the electric multipole moment, Λ will have the same behavior as the electric multipole moments.

As it was already mentioned, for the static case this whole machinery of operators and tensor analysis is not adequate to treat the multipole moments. Details are given in [23]. We therefore go straight to the dynamic case and apply the formulae.

4.2.2 Overview and Examples of the Algorithm

As a recursion procedure, we have to do the following steps to obtain the three multipolar contributions up to and including order n [23]:

1. Detracing the primitive electric multipole moment tensor of order $N = n+2$ to a symmetric traceless one using Eqs. (4.50)–(4.52).

2. Symmetrizing and detracing the magnetic multipole moment tensor of order $n + 1$ to a symmetric traceless one using first Eq. (4.47) and then Eqs. (4.50)–(4.52).

Then, the lower orders are modified to not change the fields and both steps are repeated for $n \to n - 1$. This is repeated until $n = 0$ is reached. The first step creates additional terms in the order n, the second in the orders n and $n - 1$. The primitive electric tensor of order n is changed as

$$\hat{P}^{(n)} \to \hat{\hat{P}}^{(n)} + \frac{1}{c^2}\left[\frac{n}{(n+2)^2}\dot{\hat{N}}^{(n)} - \frac{n}{2(n+2)}\ddot{\hat{\Lambda}}^{(n)}[\hat{\hat{P}}^{(n+2)}] \right] =: \hat{\hat{P}}^{(n)} + \frac{1}{c^2}\dot{\hat{T}}^{(n)} \qquad (4.53)$$

and additionally, the primitive magnetic multipole moment tensor of order $n-1$ is changed as

$$\hat{M}^{(n-1)} \to \hat{\hat{M}}^{(n-1)} + \frac{n-1}{2c^2(n+1)}\ddot{\hat{\Lambda}}^{(n-1)}[\hat{\hat{M}}^{(s,n+1)}]. \qquad (4.54)$$

The quantity $\hat{\hat{M}}^{(s,n+1)}$ is the symmetrized primitive magnetic multipole moment tensor, defined in Eq. (4.47), of order $n + 1$.

In Eq. (4.53) the toroidal multipole moment tensor of order n was defined as [54]

$$\hat{T}^{(n)}(\tau) = \frac{n}{(n+2)^2}\hat{N}^{(n)} - \frac{n}{2(n+2)}\dot{\hat{\Lambda}}^{(n)}[\hat{\hat{P}}^{(n+2)}]. \qquad (4.55)$$

In this expression for the toroidal multipole moment two terms contribute: One term with the tensor $\hat{N}$, that has, as shown above, the same parity and time inversion symmetry as we attributed to the toroidal moments in section 2.2. The second term contains the tensor $\hat{\Lambda}$. As we discussed above, $\hat{\Lambda}$ has the same space and time inversion symmetry as the electric multipole moments, but because it appears with a time derivative in the formula of $\hat{T}^{(n)}$, the time inversion symmetry of $\dot{\hat{\Lambda}}$ is the opposite of the electric parity. Thus, we have indeed found with $\hat{T}^{(n)}$ a multipole moment tensor which has the symmetry properties of the toroidal moments.

We now illustrate how the toroidal dipole and quadrupole moment can be derived using this algorithm. The toroidal dipole moment appears, like all three dipoles, in the first order of the Cartesian traceless expansion (not to be confused with the Taylor expansion in section 4.1, where it appears only in the third order). Thus, we have $n = 1$ and have to start the detracing of the primitive electric multipole moment tensor at order $N = 3$ and the symmetrization of the primitive magnetic multipole moment tensor at order $N = 2$ [54].

Step 1: Detracing the electric octupole tensor $\hat{O}^{(e)}$ using (4.52) yields

$$\Lambda_i[\hat{O}^{(e)}] = \frac{1}{5} \int (r')^2 r_i' \rho(\vec{r}',\tau) \, \mathrm{d}^3 r' \, .$$

Step 2: For $n = 1$, Eq. (4.49) yields

$$N_i = \frac{2}{3} \int [\vec{r} \times (\vec{r} \times \vec{j})]_i \, \mathrm{d}^3 r \, .$$

Now we can use Eq. (4.61) for $n = 1$,

$$\begin{aligned}
\vec{t} &= \frac{1}{4}\vec{N} - \frac{1}{6}\ddot{\vec{\Lambda}}[\hat{O}^{(e)}] \\
&= \frac{1}{6} \int [(\vec{r} \times (\vec{r} \times \vec{j}))] \, \mathrm{d}^3 r - \frac{1}{30} \int \vec{r}^2 \, \vec{r}\dot{\rho} \, \mathrm{d}^3 r \\
&= \frac{1}{10} \int [(\vec{r} \cdot \vec{j})\vec{r} - 2r^2\vec{j}] \, \mathrm{d}^3 r \, ,
\end{aligned} \tag{4.56}$$

where we again used the continuity equation (3.5) and Gauss's theorem (4.26). The same procedure is now shown for the toroidal quadrupole moment, meaning that we have to go one order higher in the detracing process.

Step 1: Detracing the electric hexadecapole tensor $\hat{H}^{(e)}$ yields

$$\Lambda_{ij}[\hat{H}^{(e)}] = \frac{1}{7} \int r^2 r_i r_j \rho(\vec{r}) \, \mathrm{d}^3 r - \frac{1}{70} \int r^4 \rho(\vec{r}) \, \mathrm{d}^3 r \, \delta_{ij} \, .$$

Step 2: The tensor $\hat{\tilde{N}}$ yields for $n = 2$

$$\tilde{N}_{ij} = \frac{3}{4} \int r_i [\vec{r} \times (\vec{r} \times \vec{j})]_j \, \mathrm{d}^3 r \, .$$

Now we can use Eq. (4.61) for $n = 2$,

$$
\begin{aligned}
Q_{ij}^{(t)} &= \frac{1}{9}\left(\tilde{N}_{ij} + \tilde{N}_{ji}\right) - \frac{1}{4}\left(\dot{\Lambda}_{ij} - \frac{1}{3}\dot{\Lambda}_{kk}\delta_{ij}\right) \\
&= \frac{1}{12}\left(\int r_i(r_j(\vec{r}\cdot\vec{j}) - j_j r^2)\,\mathrm{d}^3r + \int r_j(r_i(\vec{r}\cdot\vec{j}) - j_i r^2)\,\mathrm{d}^3r\right) \\
&\quad - \frac{1}{28}\left(\int r^2 r_i r_j \dot{\rho}(\vec{r})\,\mathrm{d}^3r - \frac{1}{3}\int r^4 \dot{\rho}(\vec{r})\,\mathrm{d}^3r\,\delta_{ij}\right) \\
&= \frac{1}{42}\int\left[4 r_i r_j(\vec{r}\cdot\vec{j}) - 5 r^2 [r_i j_j + r_j j_i] + 2 r^2 (\vec{r}\cdot\vec{j})\delta_{ij}\right]\mathrm{d}^3r
\end{aligned}
\tag{4.57}
$$

Using the traceless moments instead of the primitive ones and relation (4.53), Eq. (4.27) takes the form

$$
\vec{A}(\vec{r}, t) = \frac{\mu_0}{4\pi}\sum_{n=1}^{\infty}\frac{(-1)^{n-1}}{n!}\vec{\nabla}^{n-1}\cdot\left[\vec{\nabla}\times\frac{1}{r}\hat{M}^{(n)}(\tau) + \frac{1}{r}\dot{\hat{P}}^{(n)}(\tau) + \frac{1}{r}\frac{1}{c^2}\ddot{\hat{T}}^{(n)}(\tau)\right],
\tag{4.58}
$$

with the symmetric and traceless multipole tensors [23]

$$
\hat{P}^{(n)}(\tau) = \frac{(-1)^n}{(2n-1)!!}\int\rho(\vec{r}',\tau)(r')^{2n+1}\vec{\nabla}^n\frac{1}{r'}\,\mathrm{d}^3r',
\tag{4.59}
$$

$$
\hat{M}^{(n)}(\tau) = \frac{(-1)^{n+1}}{(n+1)(2n-1)!!}\int\sum_{k=1}^{n}(r')^{2n+1}\left(\vec{j}(\vec{r}',\tau)\times\vec{\nabla}\right)_{i_k}\prod_{\substack{m=1\\m\neq k}}^{n}\partial_{i_m}\frac{1}{r'}\,\mathrm{d}^3r'\,\vec{e}_{i_1}\otimes\cdots\otimes\vec{e}_{i_n},
\tag{4.60}
$$

$$
\hat{T}^{(n)}(\tau) = \frac{n}{(n+2)^2}\hat{N}^{(n)} - \frac{n}{2(n+2)}\dot{\hat{\Lambda}}^{(n)}.
\tag{4.61}
$$

This representation of $\vec{A}$ now actually contains the physical multipole moment tensors including the toroidal one. Now in every order n, there exist three n-th pole terms. Because of the additional curl of the magnetic multipole moment tensor, they do not appear in the same order when the vector potential is expanded with respect to r'/r. The formula for the $\hat{T}^{(n)}$ is not in a closed form, since one needs to use the electric multipole tensors to calculate $\hat{\Lambda}$. It would be nice if a closed formula could be found.

However, the various mean-square radii do not appear directly in this equation. For the mean-square radii of the electric multipole moments, this does not matter, since they do not contribute to the radiation field. But the mean-square radii of the magnetic and toroidal multipole moments do contribute, and therefore have to appear in the vector potential. When performing the detracing process for higher multipole moments than presented here, they will appear as additional terms to compensate the changes done while detracing and symmetrizing. But since we are only interested in the toroidal moments here, this procedure is not outlined in detail.

Summarizing, we showed how the toroidal moments can be deduced from the Cartesian multipole expansion by requiring definite properties of the electric and magnetic multipole moments.

However, this somehow complicated algorithm is only necessary because **the Taylor expansion, which was used as a starting point, is not an adequate approach to perform the multipole expansion**. The toroidal moment is in the above procedure from a mathematical point of view only a compensation term to leave the fields invariant when the primitive tensors are symmetrized and detraced. In the next section, we will discuss how the canonical, spherical base can be used to get the three multipole moment families in a more direct and natural way.

4.3 Multipole Expansion in Spherical Coordinates

Besides the expansion in Cartesian coordinates, one can expand the vector potential in spherical coordinates. This is based on expressing the vector potential with spherical harmonics $Y_{lm}(\theta, \phi)$ and the spherical Bessel functions $j_l(kr)$. A big advantage of the expansion in spherical coordinates is that there is no need for approximations. In Cartesian coordinates, we assumed a small source and the long-wavelength limit. This naturally limits the reliability of the calculated quantities to situations where the assumed conditions are justified. In contrast, the expansion in spherical coordinates is exact for every multipole order.

A further advantage is that the spherical multipole moments are directly traceless and symmetric and do not need to be symmetrized and detraced. We can transform them into Cartesian coordinates and vice versa by using transformation equations that will be discussed in chapter 6.

Starting with Eq. (3.22) in the Lorenz gauge and assuming the Fourier decomposition (3.8) of the current, we arrive at

$$\vec{A}_\omega(\vec{r}) = \frac{\mu_0}{4\pi} \int \vec{j}_\omega(\vec{r}') \frac{e^{ik|\vec{r}-\vec{r}'|}}{|\vec{r} - \vec{r}'|} \, d^3 r' \,.$$

We also used the dispersion relation in free space $k = \omega/c$, so that the value of k is fixed through the fact that we only consider one term of the Fourier decomposition (3.8).

We assume that the evaluation point of the potential at point $\vec{r}$ is in the exterior of the source, so that $|\vec{r}| > |\vec{r}'|$. Then, the Green's function in this integral, expressed with spherical Bessel functions, spherical Hankel functions and spherical harmonics, is [9, p. 428]

$$\frac{e^{ik|\vec{r}-\vec{r}'|}}{|\vec{r} - \vec{r}'|} = 4\pi i k \sum_{l=0}^{\infty} j_l(kr') \, h_l^{(1)}(kr) \sum_{m=-l}^{l} Y_{lm}(\theta, \phi) Y_{lm}^*(\theta', \phi') \,. \tag{4.62}$$

Explicit expressions of the functions Y_{lm} and j_l for low l are listed in the appendix. The indices l and m correspond to eigenvalues of the angular momentum operator $\mathcal{L}$, Eq. (3.23), with [55]

$$\mathcal{L}\,|l\,m\rangle = l(l+1)\,|l\,m\rangle\,, \tag{4.63}$$

$$\mathcal{L}_z\,|l\,m\rangle = m\,|l\,m\rangle\,. \tag{4.64}$$

Before discussing the dynamic case, we want to sketch the static case and identify similarities and differences between the expansion of the scalar and the vector potential.

4.3.1 The Static Case

In the static case, the expansion of the vector potential differs significantly from the expansion of the scalar potential when defining the multipole moments. For comparison, we will first show the expansion of the scalar potential, Eq. (3.21), in the static case. Therefore, we have to take the limit $k \to 0$ in Eq. (4.62), which is equivalent to assume $kr' \to 0$, since r' was anyway considered as small.

$$4\pi i k \sum_{l=0}^{\infty} j_l(kr' \to 0)\,h_l^{(1)}(kr \to 0) \sum_{m=-l}^{l} Y_{lm}(\theta,\phi)Y_{lm}^*(\theta',\phi')$$

$$= -4\pi k \sum_{l=0}^{\infty} j_l(kr' \to 0)\,n_l(kr \to 0) \sum_{m=-l}^{l} Y_{lm}(\theta,\phi)Y_{lm}^*(\theta',\phi')$$

$$= 4\pi k \lim_{k \to 0} \sum_{l=0}^{\infty} \frac{(kr')^l}{(2l+1)!!}\,\frac{(2l-1)!!}{(kr)^{l+1}} \sum_{m=-l}^{l} Y_{lm}(\theta,\phi)Y_{lm}^*(\theta',\phi')$$

$$= 4\pi \sum_{l=0}^{\infty} \frac{1}{2l+1}\frac{r'^l}{r^{l+1}} \sum_{m=-l}^{l} Y_{lm}(\theta,\phi)Y_{lm}^*(\theta',\phi')\,. \tag{4.65}$$

We used the approximation formulae for small arguments of the spherical Bessel and Hankel functions [9, p. 427], canceling the dependency of k. Using this, the static scalar potential φ becomes

$$\varphi(\vec{r}) = \frac{1}{4\pi\varepsilon_0} \int \frac{\rho(\vec{r}')}{|\vec{r}-\vec{r}'|}\,\mathrm{d}^3r'$$

$$= \frac{1}{\varepsilon_0} \sum_{l=0}^{\infty} \frac{1}{2l+1}\frac{1}{r^{l+1}} \sum_{m=-l}^{l} Y_{lm}(\theta,\phi) \int r'^l Y_{lm}^*(\theta',\phi')\rho(\vec{r}')\,\mathrm{d}^3r'$$

$$= \frac{1}{4\pi\varepsilon_0} \sum_{l=0}^{\infty} \sum_{m=-l}^{l} \sqrt{\frac{4\pi}{2l+1}}\,Y_{lm}(\theta,\phi)\frac{Q_{lm}}{r^{l+1}}\,. \tag{4.66}$$

In the last step we introduced the spherical electric multipole moment of order l [56, p. 99],

$$Q_{lm} = \sqrt{\frac{4\pi}{2l+1}} \int r'^l Y_{lm}^*(\theta',\phi')\rho(\vec{r}')\,\mathrm{d}^3r'\,. \tag{4.67}$$

This definition is made with a numeric prefactor so that for $l = 0$ the total charge and hence the Cartesian monopole moment is recovered:

$$Q_{00} = \sqrt{4\pi} \int Y_{00}^*(\theta', \phi') \rho(\vec{r}') \, d^3 r'$$
$$= \sqrt{4\pi} \int \frac{1}{\sqrt{4\pi}} \rho(\vec{r}') \, d^3 r'$$
$$= q. \tag{4.68}$$

For every index l, we get $2l+1$ independent terms ($m = -l, -l+1, ..., -1, 0, 1, ..., l-1, l$). This fits with the calculation in section 4.1.3, where we derived the number of independent entries of a traceless and symmetric Cartesian tensor and got the same result. It follows from this that the spherical multipole moments are already traceless, they do not need to be detraced. Because of the properties of the angular momentum eigenvalues l and m, the spherical multipole tensors are also symmetric.

We postpone the discussion of higher order multipole moments and the systematic conversion into Cartesian moments to chapter 6.

One could now speculate that the procedure for the static vector potential is analogous. However, this turns out to be wrong. By performing the same expansion for the vector potential as done for the scalar potential, we arrive at

$$\vec{A}(\vec{r}) = \frac{\mu_0}{4\pi} \int \frac{\vec{j}(\vec{r}')}{|\vec{r} - \vec{r}'|} \, d^3 r'$$
$$= \mu_0 \sum_{l=0}^{\infty} \frac{1}{2l+1} \frac{1}{r^{l+1}} \sum_{m=-l}^{l} Y_{lm}(\theta, \phi) \int r'^l Y_{lm}^*(\theta', \phi') \vec{j}(\vec{r}') \, d^3 r'$$
$$= \frac{\mu_0}{4\pi} \sum_{l=0}^{\infty} \sqrt{\frac{4\pi}{2l+1}} \sum_{m=-l}^{l} Y_{lm}(\theta, \phi) \frac{\vec{J}_{lm}}{r^{l+1}}, \tag{4.69}$$

where we defined the quantity

$$\vec{J}_{lm} = \sqrt{\frac{4\pi}{2l+1}} \int r'^l Y_{lm}^*(\theta', \phi') \vec{j}(\vec{r}') \, d^3 r'.$$

It is important to realize that this expansion is not a useful representation of the vector potential, since no individual term in Eq. (4.69) represents the vector potential of a specific current distribution [57]. This is due to the fact that the individual terms in Eq. (4.69) do not fulfill the properties of gauge invariance imposed on the total $\vec{A}(\vec{r})$. The index l, which appears here, does not characterize a distinct angular momentum, but each term contains a superposition of several angular momentum eigenvalues. In contrast, every term in the expansion (4.10) of the scalar potential represents a specific charge distribution [57]. Thus, the quantity $\vec{J}_{lm}$ is not a useful definition for a spherical multipole moment. The proper way how multipole moments can be defined is shown in the next section for the general dynamic case.

4.3.2 Momentum Space Formalism for the Vector Potential

We now discuss the general case for arbitrary k at the example of the vector potential.[1] Instead of the ansatz (4.69), we will decompose the current density into one longitudinal and two transverse parts.

Following Eq. (3.8), we express the current density as a Fourier series in frequency space and consider every term separately, indicated by the subscript ω at $\vec{j}$ and $\vec{A}$. Using representation (4.62), the vector potential is

$$\vec{A}_\omega(\vec{r}) = \mu_0 \mathrm{i} k \sum_{l=0}^{\infty} \sum_{n=-l}^{l} h_l^{(1)}(kr) Y_{ln}(\theta, \phi) \int \vec{j}_\omega(\vec{r}') j_l(kr') Y_{ln}^*(\theta', \phi') \, \mathrm{d}^3 r' \, . \tag{4.70}$$

We renamed the index m to n, because we will need m later for the decomposition of the current density.

We now express $\vec{j}$ through another Fourier series, this time in momentum space,

$$\vec{j}_\omega(\vec{r}') = \frac{1}{\sqrt{2\pi}} \int \tilde{\vec{j}}_\omega(\vec{\kappa}) \, \mathrm{e}^{\mathrm{i}\vec{\kappa}\cdot\vec{r}'} \, \mathrm{d}^3\kappa \, . \tag{4.71}$$

We will, as for the Fourier transformation into the frequency domain, omit the tilde on $\vec{j}$ from now on. It real space, one usually is confronted with the problem of radial integrals that tend to diverge in many cases, which enforces to limit the integration to a small spatial region. We will face this problem in the next chapter. In momentum space, however, this problem does not occur, since the integration over r can, as we will see, be performed by using a delta function. It is important to note that, because we only consider the vector potential $\vec{A}_\omega(\vec{r})$ at one distinct frequency ω, the value for k is fixed. Contrary, κ can take all values.

Similar to expansion (4.62), we represent the exponential function $\mathrm{e}^{\mathrm{i}\vec{\kappa}\cdot\vec{r}}$ as a sum of spherical harmonics and spherical Bessel functions [9, p. 471],

$$\mathrm{e}^{\mathrm{i}\vec{\kappa}\cdot\vec{r}} = 4\pi \sum_{l'=0}^{\infty} \mathrm{i}^{l'} j_{l'}(\kappa r) \sum_{m'=-l'}^{l'} Y_{l'm'}^*(\theta_\kappa, \phi_\kappa) Y_{l'm'}(\theta, \phi) \, . \tag{4.72}$$

θ and ϕ are the angles of $\vec{r}$, θ_κ and ϕ_κ characterize the angles of the vector $\vec{\kappa}$.

Now we insert Eqs. (4.71) and (4.72) into Eq. (4.70) to obtain

---

[1]The following approach in momentum space is based on personal communication with Ivan Fernandez-Corbaton.

$$\vec{A}_\omega(\vec{r}) = \frac{4\pi}{\sqrt{2\pi}}\mu_0 \mathrm{i} k \sum_{l=0}^{\infty}\sum_{n=-l}^{l} h_l^{(1)}(kr)Y_{ln}(\theta,\phi)\int \mathrm{d}^3r'\, j_l(kr')Y_{ln}^*(\theta',\phi')$$

$$\cdot \int \mathrm{d}^3\kappa\, \vec{j}_\omega(\vec{\kappa}) \sum_{l'} \mathrm{i}^{l'} j_{l'}(\kappa r') \sum_{m'=-l'}^{l'} Y_{l'm'}^*(\theta_\kappa,\phi_\kappa)Y_{l'm'}(\theta',\phi'). \tag{4.73}$$

To simplify this lengthy expression, we firstly use the orthonormality of the spherical harmonics [9, p. 108],

$$\int Y_{ln}^*(\theta',\phi')Y_{l'm'}(\theta',\phi')\,\mathrm{d}\Omega' = \delta_{ll'}\delta_{nm'}.$$

With this, the integration over θ' and ϕ' provides only non-zero values for $l = l'$ and $n = m'$. This yields

$$\vec{A}_\omega(\vec{r}) = \frac{4\pi}{\sqrt{2\pi}}\mu_0 \mathrm{i} k \sum_{l=0}^{\infty}\sum_{n=-l}^{l} h_l^{(1)}(kr)Y_{ln}(\theta,\phi)$$

$$\cdot \int r'^2\,\mathrm{d}r'\, j_l(kr') \int \mathrm{d}^3\kappa\, \vec{j}_\omega(\vec{\kappa})\, \mathrm{i}^l j_l(\kappa r')Y_{ln}^*(\theta_\kappa,\phi_\kappa). \tag{4.74}$$

Now we handle the integration over r' by using the closure relation [58]

$$\int_0^{\infty} r^2 j_l(kr)j_l(\kappa r)\,\mathrm{d}r = \frac{\pi}{2}\frac{1}{k}\frac{1}{\sqrt{\kappa k}}\delta(k-\kappa),$$

which is valid for $k,\kappa \in \mathbb{R}$. This yields

$$\vec{A}_\omega(\vec{r}) = \frac{4\pi}{\sqrt{2\pi}}\mu_0 \mathrm{i} k \sum_{l=0}^{\infty}\sum_{n=-l}^{l} h_l^{(1)}(kr)Y_{ln}(\theta,\phi)\int \mathrm{d}^3\kappa\, \vec{j}_\omega(\vec{\kappa})\, \mathrm{i}^l\, Y_{ln}^*(\theta_\kappa,\phi_\kappa)\frac{\pi}{2}\frac{1}{k}\frac{1}{\sqrt{\kappa k}}\delta(k-\kappa). \tag{4.75}$$

Finally we perform the integration over κ and we arrive at

$$\vec{A}_\omega(\vec{r}) = \frac{2\pi^2}{\sqrt{2\pi}}\mu_0 k \sum_{l=0}^{\infty}\sum_{n=-l}^{l} \mathrm{i}^{l+1} h_l^{(1)}(kr)Y_{ln}(\theta,\phi)\int \mathrm{d}\Omega_\kappa\, \vec{j}_\omega(|\vec{\kappa}|=k,\Omega_\kappa)\, Y_{ln}^*(\theta_\kappa,\phi_\kappa). \tag{4.76}$$

For the same reason as in Eq. (4.69), we will not define the integral in this equation as multipole moment. Instead, we will decompose the current density in the next section into its longitudinal and transverse fields and analyze those parts separately.

4.3.3 Decomposing the Current Density: Multipole Moments as Momentum Integrals

We have left only Fourier components of $\vec{j}_\omega$ in the expression of $\vec{A}$ which fulfill the condition $\kappa = k$. It can be shown [4] that a pure longitudinal current density with this property produces no radiation field. Thus, we can focus our discussion on the transverse part. To do this, we write $\vec{j}_\omega$ as a sum of three terms,

$$\vec{j}_\omega(\theta_\kappa, \phi_\kappa) = \sum_{j=0}^{\infty} \sum_{m=-j}^{j} \tilde{c}_{jm} \vec{W}_{jm} + \sum_{j=1}^{\infty} \sum_{m=-j}^{j} (\tilde{b}_{jm} \vec{X}_{jm} + \tilde{a}_{jm} \vec{Z}_{jm}) \,. \tag{4.77}$$

This is basically the Helmholtz decomposition (3.24). The first term represents the longitudinal part of the current, the second and third term are the two transverse parts, which generate the radiation field. The transverse parts have no term for $l = 0$, so the sum starts at $l = 1$. The prefactors $\tilde{a}_{jm}$, $\tilde{b}_{jm}$ and $\tilde{c}_{jm}$ can be calculated from the sources or from the fields. We will calculate them in the following using the current in momentum space, whereas in the next chapter, we will use both the fields itself and the sources in real space. The tilde is added to distinguish these expansion coefficients from the ones used in the next chapter. The indices l and m belong to the angular momentum of the current distribution in momentum space. Fields with different l are subject to different symmetry properties regarding rotations, whereas m is the eigenvalue of $\mathcal{L}_z$ and measures the projection of a state with total orbital angular momentum quantum number l onto the z-axis [35, p. 53]. Such a separation of the current into longitudinal and transverse parts is not Lorentz-invariant [59], however, since we are in the non-relativistic limit, there is no problem doing this decomposition.

We define the functions $\vec{W}_{jm}$, $\vec{X}_{jm}$ and $\vec{Z}_{jm}$ by using the spherical harmonics as follows [9, p. 431]:

$$\vec{W}_{jm}(\theta_\kappa, \phi_\kappa) = Y_{jm}(\theta_\kappa, \phi_\kappa)\,\vec{n}\,, \tag{4.78}$$

$$\vec{X}_{jm}(\theta_\kappa, \phi_\kappa) = \frac{1}{\sqrt{j(j+1)}} \mathcal{L}_\kappa Y_{jm}\,, \tag{4.79}$$

$$\vec{Z}_{jm}(\theta_\kappa, \phi_\kappa) = \vec{n} \times \vec{X}_{jm}\,, \tag{4.80}$$

where $\vec{n}$ is a unit vector in direction of $\vec{\kappa}$, $\vec{n} = \vec{\kappa}/\kappa$, and $\mathcal{L}_\kappa$ is the orbital angular momentum operator in momentum space, defined as

$$\mathcal{L}_\kappa = -\mathrm{i}(\vec{k} \times \vec{\nabla}_\kappa)\,.$$

The three functions $\vec{W}_{jm}$, $\vec{X}_{jm}$ and $\vec{Z}_{jm}$ are pictured in Fig. 4.2. By definition $\vec{W}$, $\vec{X}$ and $\vec{Z}$ fulfill

$$\vec{\kappa} \times \vec{W} = \vec{\kappa} \cdot \vec{X} = \vec{\kappa} \cdot \vec{Z} = 0\,,$$

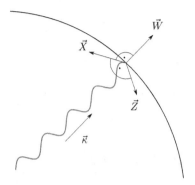

Figure 4.2: The functions $\vec{W}$, $\vec{X}$, and $\vec{Z}$, in which the current density (wavy line) is decomposed. The functions $\vec{X}$ and $\vec{Z}$ are located on the surface of a sphere and form a right angle, and $\vec{W}$ is parallel to $\vec{\kappa}$ and perpendicular to $\vec{X}$ and $\vec{Z}$ and also to the surface of the sphere.

where $\vec{\kappa}\times$ is the momentum space analogon to the real space operator $\vec{\nabla}\times$ and $\vec{\kappa}\cdot$ is the analogon to $\vec{\nabla}\cdot$. $\vec{W}$ is the longitudinal part of $\vec{j}$, and $\vec{W}$ and $\vec{Z}$ are the transverse parts. As we will show, $\vec{Z}$ accounts for the electric parity, whereas $\vec{X}$ contains the contributions from the magnetic multipole moments. Furthermore, the three functions form an orthonormal system. It holds [9, p. 472]

$$\int \vec{W}_{lm} \cdot \vec{X}_{lm} \, d\Omega = 0 \,,$$

$$\int \vec{X}_{lm} \cdot \vec{Z}_{lm} \, d\Omega = 0 \,,$$

$$\int \vec{Z}_{lm} \cdot \vec{W}_{lm} \, d\Omega = 0 \,, \tag{4.81}$$

and

$$\int \vec{W}_{jm} \cdot \vec{W}_{j'm'} \, d\Omega_\kappa = \int \vec{X}_{jm} \cdot \vec{X}_{j'm'} \, d\Omega_\kappa = \int \vec{Z}_{jm} \cdot \vec{Z}_{j'm'} \, d\Omega_\kappa = \delta_{jj'}\delta_{mm'} \,. \tag{4.82}$$

Effectively, the vector potential consists now for given indices l and j of three terms, one for each field type (electric, magnetic and longitudinal type). In real space, those terms are sometimes called "elementary vector potentials" [2]. Each of these elementary vector potentials belongs to a specific current distribution. This is the correct way to expand the vector potential we were referring to at the end of the last section.

To derive the multipole moments in momentum space, we consider in this section from now on only the transverse part, since the longitudinal part in each case does not contribute to the radiation field.

$$\vec{j}_\perp = \sum_{jm} (\tilde{b}_{jm} \vec{X}_{jm} + \tilde{a}_{jm} \vec{Z}_{jm}). \tag{4.83}$$

We now investigate the explicit expressions for the vector potential caused by some distinct indices:

$j = 1,\, l = 0$: Electric dipole term

$$\begin{aligned}
\vec{A}_{10}(\vec{r},\omega) &= \frac{2\pi^2}{\sqrt{2\pi}} \mu_0 k i h_0^{(1)}(kr) Y_{00}(\theta,\phi) \int d\Omega_\kappa \sum_{m=-1}^{1} \tilde{a}_{1m} \vec{Z}_{1m} Y_{00}^*(\theta_\kappa, \phi_\kappa) \\
&= \frac{2\pi^2}{\sqrt{2\pi}} \mu_0 k i h_0^{(1)}(kr) Y_{00}(\theta,\phi) \int d\Omega_\kappa \sum_{m=-1}^{1} \tilde{a}_{1m} \vec{n} \times \frac{1}{\sqrt{2}} \mathcal{L}_\kappa Y_{1m} Y_{00}^* \\
&= -\frac{i\mu_0\omega}{4\pi} \frac{e^{ikr}}{r} \vec{p}(k)
\end{aligned} \tag{4.84}$$

with the electric dipole moment

$$\vec{p}(k) = \frac{4\pi^2}{\sqrt{6\omega}} \begin{pmatrix} (\tilde{a}_{11} - \tilde{a}_{1-1}) \\ i\,(\tilde{a}_{11} + \tilde{a}_{1-1}) \\ -\sqrt{2}\tilde{a}_{10} \end{pmatrix}.$$

This definition of $\vec{p}(\vec{k})$ was chosen as such that the vector potential coincides with the corresponding expression in Jackson [9, p. 410]. However, this $\vec{p}(k)$ is not the same quantity as defined in Eq. (4.7). The latter equations holds only for small kr, whereas the here defined quantity is valid for all values of kr.

$j = 1,\, l = 1$: Magnetic dipole term

$$\begin{aligned}
\vec{A}_{11}(\vec{r},\omega) &= \frac{2\pi^2}{\sqrt{2\pi}} \mu_0 k \sum_{l=0}^{\infty} \sum_{n=-1}^{1} i^2 h_l^{(1)}(kr) Y_{1n}(\theta,\phi) \int d\Omega_\kappa \sum_{m=-1}^{1} \tilde{b}_{1m} \vec{X}_{1m} Y_{1m}^*(\theta_\kappa, \phi_\kappa) \tag{4.85} \\
&= \frac{2\pi^2}{\sqrt{2\pi}} \mu_0 k \sum_{l=0}^{\infty} \sum_{n=-1}^{1} i^2 h_l^{(1)}(kr) Y_{1n}(\theta,\phi) \int d\Omega_\kappa \sum_{m=-1}^{1} \tilde{b}_{1m} \frac{1}{\sqrt{2}} \mathcal{L}_\kappa Y_{1m} Y_{1m}^* \tag{4.86} \\
&= \frac{ik\mu_0}{4\pi} \frac{e^{ikr}}{r^2} \left(1 - \frac{1}{ikr}\right) \vec{r} \times \vec{m}(k) \tag{4.87}
\end{aligned}$$

with the magnetic dipole moment

$$\vec{m}(k) = \frac{\sqrt{6}\pi^2}{k} \begin{pmatrix} (\tilde{b}_{11} - \tilde{b}_{1-1}) \\ i(\tilde{b}_{1-1} + \tilde{b}_{11}) \\ -\sqrt{2}\tilde{b}_{10} \end{pmatrix}.$$

Like for the electric dipole, we defined $\vec{m}$ to fit the vector potential with Jackson's expression [9, p. 413], and like before, this definition of $\vec{m}(k)$ is the general form of Eq. (4.17), valid for all values of kr.

$\boxed{j = 1, \, l = 2: \text{Toroidal dipole term}}$

$$\vec{A}_{12}(\vec{r}, \omega) = \frac{2\pi^2}{\sqrt{2\pi}}\mu_0 k \sum_{n=-2}^{2} i^{2+1} h_2^{(1)}(kr) Y_{ln}(\theta, \phi) \int d\Omega_\kappa \sum_{m=-1}^{1} a_{1m} \vec{Z}_{1m} Y_{2m}^*(\theta_\kappa, \phi_\kappa). \tag{4.88}$$

$$= \frac{2\pi^2}{\sqrt{2\pi}}\mu_0 k \sum_{m=-2}^{2} i^{2+1} h_2^{(1)}(kr) Y_{lm}(\theta, \phi) \int d\Omega_\kappa \sum_{m=-1}^{1} \tilde{a}_{1m} \vec{n} \times \frac{1}{\sqrt{2}} \mathcal{L}_\kappa Y_{1m} Y_{2m}^* \tag{4.89}$$

$$= \frac{\mu_0}{4\pi} \frac{3 - 3ikr - k^2r^2}{r^5} (\vec{r}(\vec{r} \cdot \vec{t}(k)) - r^2\vec{t}(k)) e^{ikr} \tag{4.90}$$

with the toroidal dipole

$$\vec{t}(k) = \frac{2\pi^2}{i\sqrt{6}k^2} \begin{pmatrix} (\tilde{a}_{11} - \tilde{a}_{1-1}) \\ i(\tilde{a}_{1-1} + \tilde{a}_{11}) \\ -\sqrt{2}\tilde{a}_{10} \end{pmatrix}.$$

Like before, this definition for the toroidal moment $\vec{t}(k)$ is not the one defined in Eq. (4.45), but is valid for all values of kr. It was defined so that the expression of $\vec{A}_\omega$ matches the analogous expression found in literature [26] up to a gauge difference. Comparing to the literature expression, it holds

$$\vec{A}_{\text{lit}} - \vec{A}_\omega = \vec{\nabla}\chi$$

with

$$\vec{\nabla}\chi = \frac{e^{ikr}}{r^5} \left[2k^2r^2\vec{r}(\vec{r} \cdot \vec{t}) + (3\vec{r}(\vec{r} \cdot \vec{t}) - r^2\vec{t})(2ikr - 2) \right].$$

These terms can also be found in Eq. (4.42), if the equation is Fourier transformed into the frequency domain.

This result suggests that the electric and toroidal dipole moments are directly proportional to each other. This would implicate that by measuring one of the two moments, the other moment would also be known. Hence, **the toroidal moments seem to be an artefact of the Cartesian Taylor expansion**, where they only appear because the multipole tensors are demanded to fulfill definite properties under parity and rotations. **The fundamental character which is normally attributed to toroidal moments seems therefore to be questionable**, at least in the context of electrodynamics.

For higher values of j and l, higher multipole moments can be defined. Table 4.2 provides an overview over the multipole moments which appear in various combinations of l and j for the transverse part of $\vec{j}_\omega$. The dependence on k is suppressed. Such multipole moments depending on the wave vector k are also called "multipole form factors". Their long-wavelength limit $k \to 0$ yields the usually considered multipole moments [16].

Table 4.2: Emergence of the various multipole moments for different combinations of j and l.

j \ l	0	1	2	3
0	0	0	0	0
1	$\vec{p}$	$\vec{m}$	$\vec{t}$	0
2	0	$\hat{Q}^{(e)}$	$\hat{Q}^{(m)}$	$\hat{Q}^{(t)}$
3	0	0	$\hat{O}^{(e)}$	$\hat{O}^{(m)}$

This chapter showed that it is possible to calculate the multipoles and hence the multipole moments based on general considerations regarding the properties of the multipole tensors. For both the Cartesian and the spherical derivations, the underlying reasoning was to have tensors with definite properties. These properties concern proper transformation behavior under parity and rotations. From the spherical expansion, it is more straightforward to calculate the multipole moments because they are a priori traceless and symmetric. Contrary, in the Cartesian expansion, cumbersome transformations have to be performed until the tensors have the desired properties.

Both derivations have something in common: The toroidal moments do not appear as fundamental physical quantities. In the Cartesian expansion, they look like a mathematical necessity to compensate the detracing and symmetrizing manipulations and to leave the fields unchanged. In the spherical expansion, the toroidal dipole moment is just proportional to the electric dipole moment. We will show in the next chapter that this connection between the electric and the toroidal moments is also there when the multipole expansion is done on the level of the fields. Both multipole moments can only be disentangled when the same approximation is done like in the Cartesian Taylor expansion, namely the long-wavelength approximation.

5 Direct Multipole Expansion of Electromagnetic Fields

So far, we used the scalar and vector potential to perform the multipole expansion. In this chapter, we will show that it is also possible to perform a multipole expansion by starting directly on the level of fields. This has the advantage that the problem of the choice of gauge does not exist, since electric and magnetic fields are experimentally accessible quantities. However, in general it is not possible to reconstruct the sources unambiguously from the fields [60]. The reconstruction is only unambiguous if solely the transverse parts of the currents and of the fields are considered. This is justified here, because radiation fields only contain transverse contributions.

At first, we will decompose the electric field into its orthonormal angular momentum eigenstates with expansion coefficients to be determined. We will then discuss possibilities to perform the multipole expansion of the fields by using this decomposition. On the one hand, we will use only the fields to calculate the expansion coefficients. This will be done by projecting the total field onto the expansion terms separately, since the base functions are chosen as orthonormal. We will see that by only using this projection method, we cannot distinguish between electric and toroidal moments.

On the other hand, we will express the expansion coefficients with the charge and current densities. For this purpose, we project the fields onto a spherical surface and use the Helmholtz equation to express the projection integral with the source densities. This will also yield no clear separation between the electric and toroidal moments in the exact expression, but in the long-wavelength limit, this distinction will become clear.

5.1 Decomposition of the Fields

The electromagnetic field has one longitudinal and two transverse components. It is therefore useful to decompose the $\vec{E}$- and $\vec{B}$-fields, like the current in previous chapter, into a series of multipolar harmonic fields with complex expansion coefficients [1]. Each of these harmonic fields then will describe one of the longitudinal or transverse components. Our prefactor convention is the same as in [9].

We start with the decomposition of the electric field that reads as

$$\vec{E}(\vec{r},\omega) = Z_0 \sum_{l=0}^{\infty} \sum_{m=-l}^{l} c_{lm}\vec{L}_{lm}(\vec{r},\omega) + Z_0 \sum_{l=1}^{\infty} \sum_{m=-l}^{l} \left[b_{lm}\vec{M}_{lm}(\vec{r},\omega) + \frac{\mathrm{i}}{k}a_{lm}\vec{N}_{lm}(\vec{r},\omega) \right]. \quad (5.1)$$

The first term represents the longitudinal part of the electric field. It is related to the Coulomb field and only relevant for the near field. The discussion of this contribution can be found in section 5.4. The second and the third terms are the transverse part which belong to the radiation field [42, p. 83 et seq.]. The expansion coefficients a_{lm}, b_{lm}, and c_{lm} are related to the coefficients $\tilde{a}_{lm}$, $\tilde{b}_{lm}$, and $\tilde{c}_{lm}$ used in the previous chapter. One can show that with a suitable normalization, they are actually the same [61]. The sum indices l and m have the same meaning regarding the angular momentum as the indices j and m in the previous chapter. For the longitudinal term, the sum starts at $l = 0$, whereas for the transverse terms, there is no $l = 0$ contribution, so the sum starts at $l = 1$. In the following, we drop the arguments of the functions $\vec{E}$, $\vec{L}$, $\vec{M}$ and $\vec{N}$ to keep notation clear.

The so-called Hansen multipole harmonic fields [36] [62, p. 217] $\vec{L}$, $\vec{M}$ and $\vec{N}$ that appear in the equation above can be defined in many different ways [42, p. 87]. For example, they can be expressed trough the functions $\vec{W}(\theta_\kappa, \phi_\kappa)$, $\vec{X}(\theta_\kappa, \phi_\kappa)$ and $\vec{Z}(\theta_\kappa, \phi_\kappa)$, which we used in chapter 4 to express the current density in momentum space [63]. It holds, up to normalization factors,

$$\vec{M}_{lm}(\vec{r}) = \int \vec{X}_{lm}(\theta_\kappa, \phi_\kappa)\, e^{i\vec{\kappa}\cdot\vec{r}}\, d\Omega_\kappa$$

and analogously for $\vec{N}$ and $\vec{L}$. However, we will not use this relation. Instead, we follow [9, p. 431 et seq.] and define $\vec{L}$, $\vec{M}$ and $\vec{N}$ as follows:

$$\vec{L}_{lm} = \frac{1}{k}\vec{\nabla} j_l(kr) Y_{lm}(\theta, \phi)\,, \tag{5.2}$$

$$\vec{M}_{lm} = h_l^{(1)}(kr)\vec{X}_{lm}\,, \tag{5.3}$$

and

$$\vec{N}_{lm} = \frac{i}{k}\vec{\nabla} \times \vec{M} = \frac{i}{k}\vec{\nabla} \times h_l^{(1)}(kr)\vec{X}_{lm}\,. \tag{5.4}$$

We incorporated the natural boundary condition that requires the transverse waves to vanish at infinity by using the Hankel function of first kind and degree l, $h_l^{(1)}(kr)$ [9, p. 440]. In the longitudinal part, we use the spherical Bessel function $j_l(kr)$ to avoid a divergence at the origin. The function $\vec{X}_{lm}$ was defined in Eq. (4.79).

It is worth noting that Eq. (5.1) can also be derived by using the decomposition (3.24) of a general vector field into three Debye potentials. We mention this only for completeness and to justify Eq. (5.1). Expansion of the Debye potentials ξ, ψ, and ζ into a series of spherical harmonics and Hankel functions yields, e.g. for ξ,

$$\xi = \sum_{l=0}^{\infty} \sum_{m=-l}^{l} c_{lm} h_l^{(1)}(kr) Y_{lm}(\theta, \phi)\,.$$

Plugging these expansions into Eq. (3.24), we get, up to some arbitrary prefactors, Eq. (5.1).

Per definition $\vec{L}$, $\vec{M}$ and $\vec{N}$ fulfill

$$\vec{\nabla} \times \vec{L} = \vec{\nabla} \cdot \vec{M} = \vec{\nabla} \cdot \vec{N} = 0$$

and form, regarding the angular integration, an orthogonal system of functions. Following section 4.3.3, it holds for all l, l', m, m'

$$\int \vec{M}_{lm} \cdot \vec{N}_{l'm'} \, d\Omega = 0 \,,$$

$$\int \vec{M}_{lm} \cdot \vec{L}_{l'm'} \, d\Omega = 0 \,,$$

$$\int \vec{N}_{lm} \cdot \vec{L}_{l'm'} \, d\Omega = 0 \,. \tag{5.5}$$

Furthermore, based on the normalization of the angular part of $\vec{M}$, the same normalization holds for the angular part of $\vec{N}$:

$$
\begin{aligned}
\int \vec{N}_{lm}^* \cdot \vec{N}_{l'm'} \, d^3r &= \frac{1}{k^2} \int (\vec{\nabla} \times \vec{M}_{lm})^* \cdot (\vec{\nabla} \times \vec{M}_{l'm'}) \, d^3r \\
&= \frac{1}{k^2} \left(\int \vec{\nabla}[\vec{M}_{lm}^* \times (\vec{\nabla} \times \vec{M}_{l'm'})] \, d^3r + \int \vec{M}_{lm}^* \cdot \vec{\nabla} \times \vec{\nabla} \times \vec{M}_{l'm'} \, d^3r \right) \\
&= \frac{1}{k^2} \left(\oint [\vec{M}_{lm}^* \times (\vec{\nabla} \times \vec{M}_{l'm'})] \, d\vec{S} + \int \vec{M}_{lm}^* \cdot [\vec{\nabla}\vec{\nabla} \cdot \vec{M}_{l'm'} - \Delta\vec{M}_{l'm'}] \, d^3r \right) \\
&= -\frac{1}{k^2} \int \vec{M}_{lm}^* \cdot \Delta\vec{M}_{l'm'} \, d^3r \\
&= \int \vec{M}_{lm}^* \cdot \vec{M}_{l'm'} \, d^3r \\
&= \delta_{ll'}\delta_{mm'} \int r^2 h_l^{(1)}(kr) h_{l'}^{(1)}(kr) \, dr \,. \tag{5.6}
\end{aligned}
$$

In the second and third line we used the identities $\vec{\nabla} \cdot (\vec{A} \times \vec{B}) = \vec{B}(\vec{\nabla} \cdot \vec{A}) - \vec{A}(\vec{\nabla} \cdot \vec{B})$ and $\vec{\nabla} \times \vec{\nabla} \times \vec{A} = \vec{\nabla}\vec{\nabla} \cdot \vec{A} - \Delta\vec{A}$ valid for differentiable vector fields $\vec{A}$ and $\vec{B}$. For the third line Gauss's theorem was applied, for the fourth line it was used that the surface integral in infinity of the functions $\vec{M}$ and $\vec{N}$ vanishes as well as the divergence of $\vec{M}$, and for the last line we used the Helmholtz equation (3.12).

$\vec{L}$ is also normalized in the angular variables,

$$
\begin{aligned}
\int \vec{L}_{lm}^* \cdot \vec{L}_{l'm'} \, \mathrm{d}^3 r &= \frac{1}{k^2} \int \partial_n j_l(kr) Y_{lm}(\theta, \phi) \cdot \partial_n j_{l'}(kr) Y_{l'm'}(\theta, \phi) \, \mathrm{d}^3 r \\
&= -\frac{1}{k^2} \int j_l(kr) Y_{lm}(\theta, \phi) \, \Delta j_{l'}(kr) Y_{l'm'}(\theta, \phi) \, \mathrm{d}^3 r \\
&= \int j_l(kr) Y_{lm}(\theta, \phi) j_{l'}(kr) Y_{l'm'}(\theta, \phi) \, \mathrm{d}^3 r \\
&= \delta_{ll'} \delta_{mm'} \int r^2 j_l(kr) j_{l'}(kr) \, \mathrm{d}r \, .
\end{aligned}
\tag{5.7}
$$

In the second line we integrated by parts and in the fourth line we used the Helmholtz equation.

So we showed that $\vec{L}$, $\vec{M}$ and $\vec{N}$ form a complete base of the three dimensional vector space, and so they can be used to expand the electromagnetic field. The decomposition of the transverse part is

$$
\vec{E}_\perp = Z_0 \sum_{lm} \left[\frac{\mathrm{i}}{k} a_{lm} \vec{\nabla} \times h_l^{(1)}(kr) \vec{X}_{lm} + b_{lm} h_l^{(1)}(kr) \vec{X}_{lm} \right] ,
$$

For completeness, the corresponding magnetic field, calculated via the Maxwell equation (3.3), is given by

$$
\vec{H} = \sum_{lm} \left[a_{lm} h_l^{(1)}(kr) \vec{X}_{lm} - \frac{\mathrm{i}}{k} b_{lm} \vec{\nabla} \times h_l^{(1)}(kr) \vec{X}_{lm} \right] .
$$

The next section focuses on the assumption that the total radiation field $\vec{E}(\vec{r})$ is known, so that the coefficients a_{lm} and b_{lm} can be calculated from projecting the total field onto the base functions $\vec{M}$ and $\vec{N}$. Then, the a_{lm} and b_{lm} can, with some important restrictions, be used to calculate the various contributions of the multipole fields.

5.2 Expansion Coefficients for given Electric Field

In this section, we mainly follow works by Mühlig et al., [1] and [64], respectively. However, it should be mentioned that the results presented in these two papers are not valid when toroidal moments are present in the considered system. Improvements to the formulae given by them will be discussed subsequently.

To get values for the coefficients a_{lm} and b_{lm} in (5.1), we project the functions $\vec{N}$ and $\vec{M}$ onto the known electric field in a suitable distance $r = R_0$ from the source [1]. With suitable we mean, on the one hand, that the sphere with radius R_0 should, by any means, fully contain the sources. On the other hand, due to numerical or experimental restrictions, the sphere should be as close enough to the sources as possible. Of course, the results for a_{lm} and b_{lm} have to be independent of the choice of R_0. A suitable choice for R_0 therefore usually takes computational or experimental constraints into account [65]. Due to divergence of the radial integral, it is generally not possible to integrate till infinite radius.

The equations for calculating a_{lm} and b_{lm} in our normalization explicitely read as

$$a_{lm} = \frac{-\mathrm{i}k \int_0^{2\pi} \int_0^\pi \vec{E}(r = R_0) \cdot \vec{N}_{lm}^*(r = R_0) \sin\theta \, d\theta \, d\phi}{Z_0 \int_0^{2\pi} \int_0^\pi |\vec{N}_{lm}(r = R_0)|^2 \sin\theta \, d\theta \, d\phi} \tag{5.8}$$

and

$$b_{lm} = \frac{\int_0^{2\pi} \int_0^\pi \vec{E}(r = R_0) \cdot \vec{M}_{lm}^*(r = R_0) \sin\theta \, d\theta \, d\phi}{Z_0 \int_0^{2\pi} \int_0^\pi |\vec{M}_{lm}(r = R_0)|^2 \sin\theta \, d\theta \, d\phi}. \tag{5.9}$$

We have to find out now which of these two prefactors accounts for the electric and toroidal and which for the magnetic multipoles. We can deduce this by looking at the symmetry properties. The parity operator inverts the spatial coordinates. By using properties of the spherical harmonics, one can show that this is equivalent to [31, p. 227]

$$\mathcal{P} \, |l\,m\rangle = (-1)^{l+1} \, |l\,m\rangle \,,$$

where l and m are eigenvalues of the angular momentum operator. From this, we can deduce the following behavior:

$$\mathcal{P}\vec{M} = \frac{\mathrm{i}h_l^{(1)}(kr)}{\sqrt{l(l+1)}}(-\vec{r} \times (-\vec{\nabla}))\mathcal{P}Y_{lm} = \frac{\mathrm{i}h_l^{(1)}(kr)}{\sqrt{l(l+1)}}(\vec{r} \times \vec{\nabla})Y_{lm}(-1)^{l+1} = (-1)^{l+1}\vec{M}$$

and

$$\mathcal{P}\vec{N} = \frac{\mathrm{i}}{k}(-\vec{\nabla}) \times \vec{M}(-1)^{l+1} = (-1)^l\vec{N}\,.$$

This means that $\vec{M}$ is for $l = 1$ invariant under parity transformation, whereas $\vec{N}$ is odd. This has now to be related to the physical properties of multipoles. As discussed in chapter 2, the dipoles, meaning $l = 1$, behave like

$$\mathcal{P}\vec{p} = -\vec{p}\,, \qquad \mathcal{P}\vec{m} = \vec{m}\,.$$

From this we can conclude, since $\vec{M}$ has the same spatial symmetry as the magnetic moments, that b_{lm}, which is the amplitude coefficient in its expansion, contains the information about the magnetic moments. Consequently, $\vec{N}$ shares the spatial symmetry of the electric field and a_{lm} contains information about the multipole moments with electric parity. However, it was attributed to only electric multipole moments in [64, p. 23], herefore losing information on the toroidal contribution.

We can summarize that, though the relation between the radiating parts of the current and the radiation fields is unambiguous, the ansatz we used in Eq. (5.1) is not the appropriate one to calculate the various multipolar contributions separately. We were not able to distinguish between electric and toroidal moments because our base functions $\vec{M}$ and $\vec{N}$ reflect only different parity properties, but not the different properties under time inversion. This means that **the expansion coefficient a_{lm} in Eq. (5.8) contains both electric and toroidal contributions**, which cannot be distinguished with the preceding calculation.

In the next section, we will use the source densities to calculate the expansion coefficients. In the long-wavelength limit, this way will yield a clear distinction between electric and toroidal moments.

5.3 Expansion Coefficients for given Sources

In this section, we want to discuss how the expansion coefficients a_{lm} and b_{lm} in (5.1) can be calculated directly from the charge and current densities of the considered system. This is equivalent to using the electric and magnetic potentials for calculating the $\vec{E}$- and $\vec{B}$-fields. This path will give new insights how the coefficients a_{lm} and b_{lm} are related to the multipole moments.

The results in the last section suggested at first glance that the coefficient with the electric parity, a_{lm}, contains only the electric multipoles moments. By using the sources, we will now show that the different multipolar contributions to the expansion coefficients a_{lm} and b_{lm} can be clearly separated, at least in the long-wavelength limit [22]. The following calculation is developed in analogy to [38] and [9], but in both texts, the toroidal moments are not considered, because they are one order higher with respect to the frequency than the usual electric moments. We show some important steps for the calculation of a_{lm}, the calculation for b_{lm} is analogous.

We begin with Eq. (5.1) and project the two terms again onto a spherical surface to obtain a suitable expression for a_{lm}, but this time we multiply them first with a complex conjugated spherical harmonic function and the spatial vector $\vec{r}$. The reason for this is that we only need $\vec{r} \cdot \vec{E}_\perp$ to express the expansion coefficients through the sources, not the full $\vec{E}_\perp$ [9, p. 440]. Furthermore, we do not, contrary to the previous section, fix the radial coordinate r of the electric field. The calculation reads

$$
\begin{aligned}
\int Y_{lm}^*(\theta,\phi)\,\vec{r}\cdot\vec{E}_\perp\,\mathrm{d}\Omega &= \int Y_{lm}^*\vec{r}\cdot Z_0\sum_{l'm'}\left[b_{l'm'}\vec{M}_{l'm'}+\frac{\mathrm{i}}{k}a_{l'm'}\vec{N}_{l'm'}\right]\mathrm{d}\Omega \\
&= \int Y_{lm}^*(\theta,\phi)\,\vec{r}\cdot Z_0\sum_{l'm'}\left[\frac{\mathrm{i}}{k}a_{l'm'}\vec{\nabla}\times h_{l'}^{(1)}(kr)\vec{X}_{l'm'}\right]\mathrm{d}\Omega \\
&= \frac{\mathrm{i}}{k}Z_0 h_{l'}^{(1)}(kr)\int Y_{lm}^*\,\vec{r}\cdot\sum_{l'm'}\frac{a_{l'm'}}{\sqrt{l'(l'+1)}}\vec{\nabla}\times\mathcal{L}Y_{l'm'}\,\mathrm{d}\Omega \\
&= -\frac{Z_0}{k}h_{l'}^{(1)}(kr)\int Y_{lm}^*\sum_{l'm'}\frac{a_{l'm'}}{\sqrt{l'(l'+1)}}\mathcal{L}^2 Y_{l'm'}\,\mathrm{d}\Omega \\
&= -\frac{Z_0}{k}h_{l}^{(1)}(kr)\sqrt{l(l+1)}\,a_{lm}\,.
\end{aligned}
\tag{5.10}
$$

For the second line we used that $\vec{M}$ has no components in radial direction, therefore the scalar product with $\vec{r}$ is zero. For the third line, we took advantage of the identity $\vec{r}\cdot\vec{\nabla}\times\vec{F}=\mathrm{i}\mathcal{L}\cdot\vec{F}$, valid for every differentiable vector field $\vec{F}$. Finally, in the last line we performed the angular integration by using the orthogonality relation of the spherical harmonics and the fact that they are eigenfunctions of the squared angular momentum operator, $\mathcal{L}^2 Y_{lm}=l(l+1)Y_{lm}$.

Thus, for a_{lm} we arrive at the following equation:

$$
a_{lm} = -\frac{k}{Z_0 h_{l}^{(1)}(kr)\sqrt{l(l+1)}}\int Y_{lm}^*(\theta,\phi)\,\vec{r}\cdot\vec{E}_\perp\,\mathrm{d}\Omega\,.
\tag{5.11}
$$

In the same manner one calculates the coefficient of the magnetic contributions. Because of Maxwell's equation $\vec{\nabla}\cdot\vec{H}=0$, $\vec{H}$ is only transversal and we do not need to separate the longitudinal part.

$$
b_{lm} = \frac{k}{h_{l}^{(1)}(kr)\sqrt{l(l+1)}}\int Y_{lm}^*(\theta,\phi)\,\vec{r}\cdot\vec{H}\,\mathrm{d}\Omega\,.
\tag{5.12}
$$

To get expressions for a_{lm} and b_{lm} which depend on the sources, not on the fields, we use a path sketched in [9]. The idea is to solve Helmholtz equation (3.12), but not for the vector fields $\vec{E}$ and $\vec{H}$; instead, we apply the scalar Helmholtz equation to the scalar products $\vec{r}\cdot\vec{E}$ and $\vec{r}\cdot\vec{H}$, respectively. This is outlined in the following. It will provide an explicit expression for $\vec{r}\cdot\vec{E}_\perp$ and $\vec{r}\cdot\vec{H}$. These expressions we will plug in afterwards in (5.11) and (5.12).

As we only want to treat the transverse part of the electric field, we have to redefine it. Unlike [9], we define the divergenceless electric field as

$$
\vec{E}_\perp = \vec{E} - \frac{\vec{j}_\parallel}{\mathrm{i}\omega\varepsilon_0}\,,
\tag{5.13}
$$

so that by using the continuity equation (3.5), it holds

$$\vec{\nabla} \cdot \vec{E}_\perp = \frac{1}{\varepsilon_0}\rho - \frac{1}{i\omega\varepsilon_0}\vec{\nabla}\cdot\vec{j}_\parallel = \frac{1}{\epsilon_0}\rho - \frac{1}{i\omega\varepsilon_0}i\omega\rho = 0\,.$$

In the following, we will suppress the time dependence of $\vec{j}$ and not write it as an argument. The spatial dependence of $\vec{j}$ will only be written when it is important for the equation. Plugging the redefined electric field into Faraday's law (3.3), we get

$$\vec{\nabla} \times \vec{E}_\perp = \vec{\nabla} \times \vec{E} - \vec{\nabla} \times \frac{\vec{j}_\parallel}{i\omega\varepsilon_0} = i\omega\mu_0\vec{H} + \frac{i}{\omega\varepsilon_0}\vec{\nabla} \times \vec{j}_\parallel\,. \tag{5.14}$$

Actually, the term with $\vec{\nabla} \times \vec{j}_\parallel$ is equal to zero, but we keep it in the equations to have the full current distribution in the final results, not only the transverse part. Now we plug $\vec{E}_\perp$ into Maxwell's version of Ampère's law, Eq. (3.4),

$$\vec{\nabla} \times \vec{H} = \vec{j} + \epsilon_0\frac{\partial}{\partial t}\left(\vec{E}_\perp + \frac{\vec{j}_\parallel}{i\omega\epsilon_0}\right) = \vec{j}_\perp - i\omega\varepsilon_0\vec{E}_\perp\,. \tag{5.15}$$

We now combine Eqs. (5.14) and (5.15) to arrive at two wave equations, similar to (3.6) and (3.7), but without longitudinal field components:

$$[\Delta + k^2]\vec{E}_\perp(\vec{r},\omega) = -i\omega\mu_0\vec{j}_\perp - \frac{i}{\omega\varepsilon_0}\vec{\nabla} \times \vec{\nabla} \times \vec{j}_\parallel = -i\omega\mu_0\vec{\nabla} \times \vec{\nabla} \times \vec{j}\,, \tag{5.16}$$

$$[\Delta + k^2]\vec{H}(\vec{r},\omega) = -\vec{\nabla} \times \vec{j}\,. \tag{5.17}$$

The last step in Eq. (5.16) is outlined in the appendix. But we do not need a solution for $\vec{E}_\perp$ and $\vec{H}$, but for $\vec{r}\cdot\vec{E}_\perp$ and $\vec{r}\cdot\vec{H}$. Using several properties of the fields and relations of the orbital angular momentum operator, we are allowed to write [9, p. 429 et seq.]

$$[\Delta + k^2]\vec{r}\cdot\vec{H} = -\vec{r}\cdot(\vec{\nabla} \times \vec{j}) = -i\mathcal{L}\cdot\vec{j}\,, \tag{5.18}$$

$$[\Delta + k^2]\vec{r}\cdot\vec{E}_\perp = \frac{Z_0}{k}\vec{r}\cdot\vec{\nabla} \times (\vec{\nabla} \times \vec{j}) = \frac{Z_0}{k}\mathcal{L}\cdot(\vec{\nabla} \times \vec{j})\,, \tag{5.19}$$

where we again used the relation $\vec{r}\cdot(\vec{\nabla} \times \vec{F}) = i\mathcal{L}\cdot\vec{F}$.

The solutions for Eqs. (5.18) and (5.19) are [9, p. 440]

$$\vec{r}\cdot\vec{E}_\perp(\vec{r}) = -\frac{Z_0}{4\pi}\int\frac{e^{ik|\vec{r}-\vec{r}'|}}{|\vec{r}-\vec{r}'|}\frac{1}{k}\mathcal{L}'\cdot\vec{\nabla}_{\vec{r}'} \times \vec{j}(\vec{r}')\,\mathrm{d}^3r' \tag{5.20}$$

and

$$\vec{r} \cdot \vec{H}(\vec{r}) = \frac{i}{4\pi} \int \frac{e^{ik|\vec{r}-\vec{r}'|}}{|\vec{r}-\vec{r}'|} \mathcal{L}' \cdot \vec{j}(\vec{r}') \, d^3r' . \tag{5.21}$$

The symbol $\mathcal{L}'$ is the orbital angular momentum operator defined in (3.23), but acting on $\vec{r}'$ instead of $\vec{r}$.

We now plug Eqs. (5.20) and (5.21) into Eqs. (5.11) and (5.12) and use [9, p. 440]

$$\frac{1}{4\pi} \int d\Omega \, Y_{lm}^*(\theta, \phi) \frac{e^{ik|\vec{r}-\vec{r}'|}}{|\vec{r}-\vec{r}'|} = ik h_l^{(1)}(kr) j_l(kr') Y_{lm}^*(\theta', \phi') .$$

This yields for a_{lm}

$$
\begin{aligned}
a_{lm} &= -\frac{1}{Z_0 h_l^{(1)}(kr)} \frac{k}{\sqrt{l(l+1)}} \int Y_{lm}^*(\theta, \phi) \, \vec{r} \cdot \vec{E}_\perp \, d\Omega \\
&= -\frac{1}{Z_0 h_l^{(1)}(kr)} \frac{k}{\sqrt{l(l+1)}} \int Y_{lm}^*(\theta, \phi) \left(-\frac{Z_0 k}{4\pi} \int \frac{e^{ik|\vec{r}-\vec{r}'|}}{|\vec{r}-\vec{r}'|} \frac{1}{k^2} \mathcal{L}' \cdot \vec{\nabla}_{\vec{r}'} \times \vec{j}(\vec{r}') \, d^3r' \right) d\Omega \\
&= \frac{1}{Z_0 h_l^{(1)}(kr)} \frac{Z_0 k}{k^2} \frac{k}{\sqrt{l(l+1)}} \int \mathcal{L}' \cdot \vec{\nabla}_{\vec{r}'} \times \vec{j}(\vec{r}') \, ik \, h_l^{(1)}(kr) j_l(kr') Y_{lm}^*(\theta', \phi') \, d^3r' \\
&= \frac{ik}{\sqrt{l(l+1)}} \int j_l(kr) Y_{lm}^*(\theta, \phi) \, \mathcal{L} \cdot \vec{\nabla} \times \vec{j}(\vec{r}) \, d^3r \tag{5.22}
\end{aligned}
$$

and the same calculation for b_{lm} results in

$$b_{lm} = -\frac{k^2}{\sqrt{l(l+1)}} \int j_l(kr) Y_{lm}^*(\theta, \phi) \mathcal{L} \cdot \vec{j}(\vec{r}) \, d^3r . \tag{5.23}$$

After some algebraic transformations [9, p. 440 et seq.], we arrive at the following expressions for a_{lm} and b_{lm}:

$$a_{lm} = \frac{k}{\sqrt{l(l+1)}} \int Y_{lm}^* \left[-i\omega\rho \frac{\partial}{\partial r}[r j_l(kr)] + k^2(\vec{r} \cdot \vec{j}) j_l(kr) \right] d^3r \tag{5.24}$$

and

$$b_{lm} = -\frac{ik^2}{\sqrt{l(l+1)}} \int j_l(kr) Y_{lm}^*(\theta, \phi) \nabla \cdot (\vec{r} \times \vec{j}) \, d^3r . \tag{5.25}$$

(5.24) consists of two terms: One term with the charge density ρ, and one term where only the radial components of $\vec{j}$ appear. Note that we only used the transverse part of the current density, thus the charge density should, strictly speaking, not occur. However, it can be added as a phantom term in this derivation, where it is just zero. Slightly different derivations [9] show that the term containing the charge density is present.

In this form, the two results for a_{lm} and b_{lm} look quite awkward. However, they can be simplified in the long-wavelength limit, meaning that $kr \ll 1$. Under this assumption, the Bessel function can be approximated with [9, p. 427]

$$j_l(x) \simeq \frac{x^l}{(2l+1)!!} \left(1 - \frac{x^2}{2(2l+3)} + ... \right).$$

This approximation can be truncated after arbitrary terms, depending on the desired accuracy. In [9, p. 441], only the zeroth order contribution in (5.24) was kept and the first omitted. However, as it was pointed out in 2002 [22], the first order approximation (thus keeping terms with k^2) yields the toroidal moments. To show this for the dipoles, we set $l = 1$ and expand the Bessel function in the first term of a_{lm} until the first order, but for the Bessel function in the second term of a_{lm} we keep only the zeroth order, so that in both terms the highest order of k is k^4.

$$\begin{aligned}
a_{1m} &= \frac{k}{\sqrt{2}} \int Y_{1m}^* \left[-i\omega\rho \frac{\partial}{\partial r}[rj_1(kr)] + k^2(\vec{r}\cdot\vec{j})j_1(kr) \right] \mathrm{d}^3r \\
&\simeq \frac{k}{\sqrt{2}} \int Y_{1m}^* \left[-i\omega\rho \left[\left(\frac{2kr}{3} - \frac{4k^3r^3}{30} \right) \right] + k^2(\vec{r}\cdot\vec{j}) \left(\frac{kr}{3} \right) \right] \mathrm{d}^3r \qquad (5.26)
\end{aligned}$$

Some manipulations when plugging in $m = \{-1, 0, 1\}$ can be done using the continuity equation (3.5) and Gauss's theorem (4.26). When combining the different a_{1m} in a suitable way, we arrive at the relation [22]

$$\begin{pmatrix} a_{11} - a_{1-1} \\ \mathrm{i}(a_{11} + a_{1-1}) \\ -\sqrt{2}a_{10} \end{pmatrix} \simeq -\frac{k^2}{\sqrt{3\pi}}(-i\omega\vec{p} + k^2\vec{t}), \qquad (5.27)$$

with the well-known definitions for the electric and toroidal dipoles, Eqs. (4.7) and (4.45). Thus, the contributions of the electric and toroidal dipoles have been disentangled in the long-wavelength limit.

Equation (5.27) can also be given for arbitrary l and m as well as arbitrary high orders of approximations of the Bessel function [22]. It holds

$$a_{lm} = -\frac{k^2}{\sqrt{3\pi}}[-i\omega Q_{lm}(0, \omega) + k^2 T_{lm}(-k^2, \omega)], \qquad (5.28)$$

where $Q_{lm}(0,\omega)$ are the spherical electric multipole moments defined in section 4.3 and $T_{lm}(-k^2,\omega)$ is the toroidal multipole form factor [16]. The long-wavelength limit of $T_{lm}(-k^2,\omega)$ leads to the spherical toroidal multipole moment $T_{lm}(0,\omega)$ with

$$T_{lm}(0,\omega) = -\frac{\sqrt{\pi l}}{2l+1} \int r^{l+1} \left[\vec{Y}_{l\,l-1m}^*(\theta,\phi) + \frac{2\sqrt{l/(l+1)}}{2l+3} \vec{Y}_{l\,l+1m}^*(\theta,\phi) \right] \cdot \vec{j}_\omega(\vec{r})\, \mathrm{d}^3 r. \tag{5.29}$$

The definitions of the vector spherical harmonics $\vec{Y}_{ll'm}^*$ are given in the appendix.

Considering the b_{lm} and keeping only the zeroth order of the Bessel function, we get

$$
\begin{aligned}
b_{lm} &= -\frac{k^2}{\sqrt{l(l+1)}} \int j_l(kr) Y_{lm}^*(\theta,\phi)\mathcal{L}\cdot\vec{j}(\vec{r})\,\mathrm{d}^3 r \\
&\simeq \frac{k^2}{\sqrt{l(l+1)}} \frac{1}{(2l+1)!!} \int r^l \vec{j}(\vec{r}) \cdot \mathcal{L} Y_{lm}^*(\theta,\phi)\,\mathrm{d}^3 r \\
&= \frac{-ik^2}{\sqrt{l(l+1)}} \frac{1}{(2l+1)!!} \int r^l \vec{j}(\vec{r}) \cdot (\vec{r}\times\vec{\nabla}) Y_{lm}^*(\theta,\phi)\,\mathrm{d}^3 r \\
&= \frac{-ik^2}{(2l+1)!!} \sqrt{\frac{2l+1}{4\pi}} M_{lm}
\end{aligned}
\tag{5.30}
$$

with the usual spherical magnetic multipole moments [38]

$$M_{lm} = \frac{1}{\sqrt{l(l+1)}} \sqrt{\frac{4\pi}{2l+1}} \int r^l \vec{j}(\vec{r}) \cdot (\vec{r}\times\vec{\nabla}) Y_{lm}^*(\theta,\phi)\,\mathrm{d}^3 r. \tag{5.31}$$

In the coefficient b_{lm}, there is no contribution from another multipole family, because, as already mentioned in chapter 2, such a multipole family does not exist in Maxwell-Lorentz electrodynamics.

We have thus shown a connection between the exact multipole moments and their long-wavelength limit [66]. We could demonstrate that the toroidal moments appear as the next term after the electric dipole moment when the exact expansion coefficient a_{lm} is expanded for small values of the wave number.

5.4 Physical Relevance of the Longitudinal Part

Although not treated in [61] and [9], for a complete discussion of the scattered electric field we need to take into account its longitudinal part and wish to provide an expansion for it. It is given by the first term of Eq. (5.1),

$$\vec{E}_\parallel = Z_0 \sum_{lm} \left[c_{lm}\, j_l(kr)\vec{W}_{lm} \right] = \frac{Z_0}{k} \sum_{lm} \left[c_{lm}\, \vec{\nabla} j_l(kr) Y_{lm}(\theta,\phi) \right]. \tag{5.32}$$

From the definition of the transverse part of $\vec{E}$, Eq. (5.13), we get another definition of $\vec{E}_\parallel$ [35, p. 15]:

$$\vec{E}_\parallel = \frac{\vec{j}_\parallel}{i\omega\varepsilon_0} \, . \tag{5.33}$$

So we can equate Eqs. (5.32) and (5.33) and take the divergence of it,

$$
\begin{aligned}
\vec{\nabla} \cdot \vec{E}_\parallel &= \frac{Z_0}{k} \sum_{lm} [c_{lm} \, \Delta j_l(kr) Y_{lm}(\theta,\phi)] \\
&= -Z_0 \sum_{lm} [c_{lm} \, k j_l(kr) Y_{lm}(\theta,\phi)] \\
&= \frac{1}{\varepsilon_0} \rho \, .
\end{aligned}
\tag{5.34}
$$

Here we used in the second line the Helmholtz equation (3.25) and in the third line Gauss's law (3.1).

We now project Eq. (5.34) on a spherical surface by multiplying it with $Y_{lm}^*(\theta,\phi) j_l(kr)$. To avoid divergence of the radial integral, we just integrate until a radius R_0, that must be big enough so that the corresponding sphere encloses the sources completely.

$$
\begin{aligned}
\int Y_{lm}^*(\theta,\phi) j_l(kr) \frac{\rho}{\varepsilon_0 Z_0 k} \, \mathrm{d}^3 r &= -\int j_l(kr) Y_{lm}^*(\theta,\phi) \sum_{l'm'} c_{l'm'} j_{l'}(kr) Y_{l'm'}(\theta,\phi) \, \mathrm{d}^3 r \\
&= -\int_0^{R_0} r^2 j_l(kr) \sum_{l'm'} c_{l'm'} j_{l'}(kr) \delta_{ll'} \delta_{mm'} \, \mathrm{d}r \\
&= -c_{lm} \int_0^{R_0} r^2 j_l(kr) j_l(kr) \, \mathrm{d}r \, .
\end{aligned}
\tag{5.35}
$$

The integral $\int_0^{R_0} r^2 j_l(kr) j_l(kr) \mathrm{d}r$ is just a number, and it follows

$$c_{lm} = -\frac{1}{\int_0^{R_0} r^2 j_l(kr) j_l(kr) \, \mathrm{d}r} \int Y_{lm}^*(\theta,\phi) j_l(kr) \frac{\rho}{\varepsilon_0 Z_0 k} \, \mathrm{d}^3 r \, .$$

As we see, the coefficient c_{lm} for the longitudinal part of the electromagnetic field is closely related to the decomposition (5.34) of the charge density ρ into spherical Bessel functions and spherical harmonics. This ties with the statement [35, p. 16] that the longitudinal electric field is basically the Coulomb field, when the charge density ρ is assumed as static and the propagation velocity as infinite. Also, the longitudinal part of the vector potential, which arises from the longitudinal part of the current, is a consequence of gauge and has no physical meaning [59]. It is therefore sufficient to only consider the transverse part of the current and of the fields when dealing with the radiation field.

6 Connection and Comparison of the Different Approaches

In the previous chapters, we analyzed the multipole expansions starting both from the potentials and from the fields. We found that a comprehensive derivation of the toroidal moments is possible on the level of the sources. In the expressions of the potentials we introduced the concept of mean-square radii, which can also have effects on the fields. From symmetry arguments, we could deduce that the fields of the electric and toroidal moments look almost identically. Later in this chapter we point to their differences and discuss physical implications. Now, we wish to compare the various approaches for the multipole expansion. We give some formulae to transform the spherical multipole moments into the Cartesian multipole moments. For the dipole order, this is done explicitly, for higher orders transformation formulae are given. Furthermore, we will compare the physical relevant $\vec{E}$- and $\vec{B}$-fields as caused by the multipole moments of interest. This completely avoids gauge issues and terms in the potentials which have no effect on the fields.

6.1 Conversion of Spherical into Cartesian Moments and Vice Versa

Generally, it can be a disadvantage of the spherical multipole moments Q_{lm} and M_{lm} as well as of the coefficients a_{lm} and b_{lm} that they characterize the multipole contributions in spherical coordinates. For practical purposes it is often convenient to use Cartesian coordinates. Therefore, transformation rules have to be found to express the Cartesian multipole moments in terms of the spherical multipole moments. Thus, in this section we want to demonstrate the transformation for the dipole order and provide equations which can be used to perform the transformation for arbitrary orders.

Before going into the discussion of the transformation formulae, we clarify the notation. We use $\vec{p}$, $\vec{m}$, $\hat{Q}^{(e)}$, $\hat{Q}^{(m)}$ etc. for the Cartesian electric and magnetic multipole moments derived in section 4.2. Because of the approximations done in the Taylor series, these multipole moments are only valid for small sources in the long-wavelength limit. In Eq. (4.67) we defined for the static case the spherical electric multipole moments Q_{lm} and in Eq. (5.31) the static magnetic spherical multipole moment M_{lm}. The a_{lm}, $\tilde{a}_{lm}$ b_{lm}, $\tilde{b}_{lm}$ from chapters 4 and 5 are valid for the dynamic case and for an arbitrary extent of the source. We showed that in the long-wavelength limit, they can be expressed through the quantities Q_{lm} and M_{lm}. Keeping this in mind, we only discuss transformation formulae for the Q_{lm} and M_{lm}, since the a_{lm} and b_{lm} transform in the same way.

In section 4.3.1, we haven chosen the normalization constant of the spherical multipole moments in a way that for $l = 0$ the monopole moment is exactly reached, $Q_{00} = q$. We now look at the $l = 1$ terms:

$$Q_{10} = \int \rho\, r \cos\theta\, \mathrm{d}^3 r$$
$$= \int \rho\, z\, \mathrm{d}^3 r \equiv p_z , \tag{6.1}$$

where we used $z = r\cos\theta$ and the definition of the electric dipole in Eq. (4.7). In the same manner one calculates [22]

$$Q_{11} = -\frac{1}{\sqrt{2}} \int \rho\, r \sin\theta e^{-\mathrm{i}\phi}\, \mathrm{d}^3 r$$
$$= -\frac{1}{\sqrt{2}} \int \rho\, r \sin\theta (\cos\phi - \mathrm{i}\sin\phi)\, \mathrm{d}^3 r$$
$$= -\frac{1}{\sqrt{2}} \int \rho\, (x - \mathrm{i}y)\, \mathrm{d}^3 r$$
$$\equiv \frac{1}{\sqrt{2}} (\mathrm{i}p_y - p_x) \tag{6.2}$$

and analogous for Q_{1-1}

$$Q_{1-1} = \frac{1}{\sqrt{2}} (\mathrm{i}p_y + p_x) .$$

Inverting these relations provides the dependence of the Cartesian moments on the spherical moments,

$$p_x = -\frac{1}{\sqrt{2}} (Q_{11} + Q_{11}^*) \quad \text{or} \quad p_y = \frac{\mathrm{i}}{\sqrt{2}} (Q_{11}^* - Q_{11}) \tag{6.3}$$

and

$$p_x = \frac{1}{\sqrt{2}} (Q_{1-1} - Q_{11}) \quad \text{or} \quad p_y = -\frac{\mathrm{i}}{\sqrt{2}} (Q_{11} + Q_{1-1}) . \tag{6.4}$$

So in total we can write

$$\vec{p} = -\frac{1}{\sqrt{2}} \begin{pmatrix} (Q_{11} - Q_{1-1}) \\ \mathrm{i}(Q_{11} + Q_{1-1}) \\ -\sqrt{2}Q_{10} \end{pmatrix} . \tag{6.5}$$

For the a_{lm}, such a formula containing also spherical toroidal moments, was already given in Eq. (5.27). In analogy, the Cartesian magnetic dipole moment can be expressed through the spherical magnetic multipole moments, defined in (5.31), as

$$\vec{m} = -\frac{1}{\sqrt{2}} \begin{pmatrix} (M_{11} - M_{1-1}) \\ i(M_{11} + M_{1-1}) \\ -\sqrt{2}M_{10} \end{pmatrix}. \tag{6.6}$$

Using the spherical toroidal multipole moments, Eq. (5.29), the Cartesian toroidal dipole moment is

$$\vec{t} = -\frac{1}{\sqrt{2}} \begin{pmatrix} (T_{11} - T_{1-1}) \\ i(T_{11} + T_{1-1}) \\ -\sqrt{2}T_{10} \end{pmatrix}. \tag{6.7}$$

In [22] and [27] formulae are given to transform the multipole moments up to third order from spherical into Cartesian base. These formulae were supposedly calculated for every order separately. It is more convenient to use a general formula, where arbitrary high orders can be calculated. Such a formula for general Cartesian and spherical tensors was given in [67]. The entries $(T_l)_{pqr}$ of a Cartesian tensor of rank $p + q + r = l$ can be calculated from the spherical tensor components T_{lm} with the formula

$$(T_l)_{pqr} = \frac{i^q}{2^{p+q}} \sum_{\substack{m=0 \\ p+q+m \text{ even}}}^{p+q} C(p,q,m) A_{lm}^{(-1)^p},$$

where A_{lm}^{ϵ} with $\epsilon = (-1)^p = \pm 1$ is given through

$$A_{lm}^{\epsilon} = \frac{1}{1+\delta_{m0}} \left(\frac{2^l (l-m)!(l+m)!}{(2l)!} \right)^{\frac{1}{2}} [T_{lm} + \epsilon T_{l-m}].$$

The numerical prefactor $C(p,q,m)$ reads as

$$C(p,q,m) = \frac{q!}{[1/2(p+q+m)]! \, [1/2(q-p-m)]!} \, {}_1F_2[-p, -1/2(p+q+m); 1/2(q-p-m)+1; -1],$$

where ${}_1F_2$ is the hypergeometric function. p, q, and r are the exponents of the coordinates x, y, and z. E.g. for $p = 1$, $q = 3$, and $r = 2$, we have $(T_l)_{pqr} = (T_6)_{132} = xy^3z^2$.

These formulae produce exactly the relations given in Eqs. (6.2)–(6.7). The formulae to go in the other direction, from Cartesian to spherical multipole moments, can be also found in [67]. We do not discuss it here, since our discussion of the fields hereafter uses only Cartesian moments.

6.2 Properties of the Fields including Toroidal Moments

So far, we have calculated several scalar and vector potential expressions. The important physical quantities, however, are the electric and magnetic fields. Thus, in this section we will discuss the fields $\vec{E}$ and $\vec{B}$ originating from the different approaches of the multipole expansions.

As Raab and Lange pointed out [10, p. 27 et. seq.], it is in general not sufficient to only consider the traceless electric and magnetic multipole moment tensors. When comparing the vector potential in first order for the two multipole moment definitions, namely primitive [Eq. (4.22)] and traceless [Eq. (4.40)], we see a discrepancy, but the electric and magnetic fields for both expressions of the vector potential will be the same. This is because the mean-square radius of the electric dipole does not, as shown before, contribute to the fields. The electric far field originating from the zeroth and first order Cartesian Taylor expansion of the vector potential, no matter if Eq. (4.22) or Eq. (4.40) is used, is [68]

$$\vec{E}^{(0+1)}(\vec{r},\omega) = \frac{k^2}{4\pi\epsilon_0}\frac{e^{ikr}}{r}\left\{\vec{n}\times(\vec{p}\times n) + \frac{1}{c}(\vec{m}\times\vec{n}) + \frac{ik}{2}\vec{n}\times[\vec{n}\times(\hat{Q}^{(e)}\cdot\vec{n})]\right\}, \qquad (6.8)$$

where $\vec{n} = \vec{r}/r$ and only terms $\sim 1/r$ are kept. From this formula for $\vec{E}$, the magnetic field, if needed, could be calculatad via

$$\vec{B} = \frac{1}{c}\vec{n}\times\vec{E}.$$

For the third order, there is a discrepancy when using primitive or traceless moments. The electric field in second order with primitive moments, calculated from Eq. (4.23), is

$$\vec{E}^{(2)}_{\text{prim}}(\vec{r},\omega) = \frac{k^2}{4\pi\epsilon_0}\frac{e^{ikr}}{r}\left\{\frac{ik}{2c}\vec{n}\times(\hat{\hat{Q}}^{(m)}\cdot\vec{n}) + \frac{k^2}{6}\vec{n}\times[\vec{n}\times(\hat{\hat{O}}^{(e)}\cdot\vec{n})\cdot\vec{n}]\right\}, \qquad (6.9)$$

whereas for traceless moments (and neglecting the terms arising in the symmetrizing and detracing process), this would be

$$\vec{E}^{(2)}_{\text{tra-less}}(\vec{r},\omega) = \frac{k^2}{4\pi\epsilon_0}\frac{e^{ikr}}{r}\left\{\frac{ik}{2c}\vec{n}\times(\hat{Q}^{(m)}\cdot\vec{n}) + \frac{k^2}{6}\vec{n}\times[\vec{n}\times(\hat{O}^{(e)}\cdot\vec{n})\cdot\vec{n}]\right\}. \qquad (6.10)$$

One can show that in general $\vec{E}^{(2)}_{\text{prim}} \neq \vec{E}^{(2)}_{\text{tra-less}}$ [10, p. 27], because in general $\hat{\hat{O}}^{(e)} \neq \hat{O}^{(e)}$ and $\hat{\hat{Q}}^{(m)} \neq \hat{Q}^{(m)}$. We therefore have to take into account the contributions from symmetrizing and detracing the multipole tensors and use Eq. (4.42). This yields for the three lowest orders [68]

$$\vec{E}^{(0+1+2)}(\vec{r}, \omega) = \frac{k^2}{4\pi\epsilon_0} \frac{e^{ikr}}{r} \left\{ \vec{n} \times (\vec{p} + ik\vec{t}) \times n + \frac{1}{c}(\vec{m} \times \vec{n}) + \frac{ik}{2}\vec{n} \times [\vec{n} \times (\hat{Q}^{(e)} \cdot \vec{n})] \right.$$
$$\left. + \frac{ik}{2c}\vec{n} \times (\hat{Q}^{(m)} \cdot \vec{n}) + \frac{k^2}{6}\vec{n} \times [\vec{n} \times (\hat{O}^{(e)} \cdot \vec{n}) \cdot \vec{n}] \right\}. \tag{6.11}$$

We are therefore urged to include the toroidal dipole moment if we want to describe the electric field up to the second order properly. For higher orders, this of course is also true, and only in the zeroth and first order, a toroidal moment does not occur.

We now turn to the exact fields of a toroidal dipole. As already mentioned in chapter 2, the electric and magnetic fields of a toroidal dipole are almost the same as the fields of an electric dipole. The fields of a toroidal dipole read as [26]

$$\vec{E}^{(t)}(\vec{r}, \omega) = \frac{1}{4\pi\epsilon_0} \left(\frac{ik^3}{r}(\vec{n} \times \vec{t}) \times \vec{n} + \left(\frac{ik}{r^3} + \frac{k^2}{r^2} \right) \left[3\vec{n}(\vec{n} \cdot \vec{t}) - \vec{t} \right] \right) e^{ikr} \tag{6.12}$$

and

$$\vec{B}^{(t)}(\vec{r}, \omega) = \frac{c}{4\pi} k^2 (\vec{n} \times \vec{t}) \left(\frac{ik}{r} - \frac{1}{r^2} \right) e^{ikr}. \tag{6.13}$$

In comparison, the fields of an electric dipole are [9, p. 411]

$$\vec{E}^{(p)}(\vec{r}, \omega) = \frac{1}{4\pi\epsilon_0} \left(\frac{k^2}{r}(\vec{n} \times \vec{p}) \times \vec{n} + \left(\frac{1}{r^3} - \frac{ik}{r^2} \right) [3\vec{n}(\vec{n} \cdot \vec{p}) - \vec{p}] \right) e^{ikr} \tag{6.14}$$

and

$$\vec{B}^{(p)}(\vec{r}, \omega) = \frac{c}{4\pi} k^2 (\vec{n} \times \vec{p}) \left(\frac{1}{r} - \frac{1}{ikr^2} \right) e^{ikr}. \tag{6.15}$$

Thus, the fields of electric and toroidal dipole (and in general all fields of the electric and toroidal n-th pole) only differ in the factor ik. This has two implications: First, the radiated intensity of a toroidal dipole is scaled with the factor k^2 compared to the intensity of the electric dipole (under the assumption that $\vec{p}$ and $\vec{t}$ do not depend on ω). It is known that the radiated intensity of an electric dipole has the shape of the Lorentz curve [42, p. 230]. This means that by measuring the radiated intensity for different frequencies, one should be able to identify if the dipole is of electric or toroidal kind [22].

Secondly, the imaginary unit i generates a phase-shift of $\pi/2$ of the toroidal dipole field with respect to the electric dipole field. This could be exploited to design a source distribution where the electric and toroidal dipole annihilate each other. The sum of both fields is

$$\vec{E}^{(tot)} = \vec{E}^{(p)} + \vec{E}^{(t)}$$

$$= \frac{1}{4\pi\epsilon_0} \left(\frac{k^2}{r} (\vec{n} \times (\vec{p} + ik\vec{t}) \times \vec{n} + \left(\frac{1}{r^3} - \frac{ik}{r^2} \right) [3\vec{n}(\vec{n} \cdot (\vec{p} + ik\vec{t})) - (\vec{p} + ik\vec{t})] \right) e^{ikr}. \quad (6.16)$$

Thus, if we design a charge-current distribution in a way that $\vec{p} = -ik\vec{t}$, the electric and magnetic fields vanish both in the near and in the far field [26]. Of course this is only possible if $\vec{p}$ and $\vec{t}$ can be tuned independently from each other, which is, considering the results of the previous chapter, not possible. This destructive interference would be then the anapole configuration described in section 2.4. The vector potential, however, does not vanish, enabling Aharonov-Bohm like effects [19]. It was reported recently [7] that a non-radiating current-distribution was found in dielectric nanoparticles and interpreted as an anapole. Also it seems possible to make a nanowire invisible by interference of electric and toroidal dipoles [8].

7 Summary and Outlook

The research field of toroidal multipole moments has been of growing interest in recent years. This thesis should provide a contribution to this research by discussing the relevance of toroidal moments and possibilities how they can be included in the classical electromagnetic multipole expansions.

Chapter 2 contained the motivation, why toroidal moments are a useful concept for describing certain current distributions. We gave an overview of space and time symmetry properties of the four possible dipole families. The discussion was then limited to three multipole families because the forth, the axial toroidal moments, are usually considered as not realizable in elementary charge-current distributions.

In chapter 4 we analyzed various multipole expansions on the level of the potentials and discussed several possibilities how the toroidal moments can be included naturally into these formalisms. Starting from the multipole expansion of the vector potential, we outlined that with the Cartesian Taylor expansion it is not feasible to derive the toroidal moments. Such Taylor expansion however is the usual approach taught in the context of lectures on electrodynamics, which might explain why toroidal multipole moments have not been considered. The multipole tensors shall have definite properties under rotations and parity. The multipole tensors emerging from the Taylor expansion do not have such definite properties and are thus not appropriate multipole tensors. By using certain prescriptions to identify the relevant contributions of the tensors, the toroidal dipole moment could be identified. However, it is practically not feasible to decompose those tensors for each higher order separately into their fundamental constituents, which then would have definite properties.

A more algorithmic approach asks to translate the requirement of definite parity and rotational properties into the mathematical conditions of tensor symmetry and tracelessness. We discussed such an algorithm with which it is possible to calculate the toroidal moments for arbitrary high orders. Hereby, we showed that the toroidal multipole moments have contributions from detracing the electric multipole tensors as well as from symmetrizing and detracing the magnetic multipole tensors. We derived a form of the vector potential containing the three distinct multipolar moments. This expression, Eq. (4.58), and the formulae (4.59)–(4.61) enable the calculation of the Cartesian multipole moments for arbitrary order. However, the expression for the toroidal multipole tensor is not self-contained, because one needs to know the traces of the two orders higher electric multipole tensor. Generally, it would be nice to have an expression for the toroidal multipole tensor which does not rely on the electric dipole tensor.

We then turned to the expansion of the vector potential in spherical coordinates. As a difference to the previous Cartesian derivations, we carried out the calculations with the current density in momentum space and separated the longitudinal and transverse parts. This derivation is exact and does not contain, contrary to the Cartesian expansion, the approximation of small sources in the long-wavelength limit. We showed that by decomposing the current density in momentum space into its transverse and longitudinal parts, we were able to derive expressions for the various multipole moments. Comparing these expressions suggested that the electric and toroidal dipoles are proportional to each other for a given current distribution. This is an important finding since it suggests that the electric and toroidal multipole moments cannot be adjusted independently. They are intimately linked. Once the electric dipole moment is fixed for a given vector potential, the toroidal multipole moment is given as well. The implication of this finding in the context of metamaterials still needs to be discussed, because it has been suggested that the multipole moments can be accessed somehow independently from each other [26].

In chapter 5, we performed the multipole expansion on the level of the fields. Similar to the current density in chapter 4, we decomposed the electric and magnetic fields into its longitudinal and transverse parts. We argued that the longitudinal part is not relevant for the radiation field, and hence focused only on the transverse part. Following this, we outlined two possibilities to calculate the expansion coefficients of the field decomposition. The one which only uses the fields itself mixes up the contributions from electric and toroidal moments. The distinction between electric and toroidal moments could only be restored when we used the sources to express the expansion coefficients and took the long-wavelength limit. This is very helpful since it provides a link between the results obtained with the Taylor expansion in Cartesian coordinates and the expansion in spherical coordinates.

In chapter 6 we compared the different approaches and linked the Cartesian multipole moments to the multipole moments in spherical coordinates. We discussed the electric fields of primitive and traceless multipole tensors and outlined differences between the two approaches. Furthermore, we compared the exact fields of an electric dipole with those of a toroidal dipole and showed how an anapole can be generated.

Summarizing, we analyzed the electromagnetic multipole expansion with respect to the toroidal moments, and we found that their emergence is closely related to the long-wavelength approximation. Only be demanding definite properties of the multipole tensors, the toroidal moments appear as part of the Cartesian Taylor expansion. This may be a reason why they are overlooked often when doing the multipole expansion. Our calculations in spherical coordinates suggest that the electric and toroidal multipoles are both built up from the same combination of coefficients, and differ only in the prefactor. This would make it impossible to tune both dipoles independently, and toroidal moments would always be present in a system which has an electric moment.

Even though many important questions have been answered, a few open questions remain. First of all, an experiment including the feature of time inversion should be carried out. This would provide a clear indication of toroidal moments and their distinction to electric moments.

Another question, which was not addressed in this thesis, concerns the physical realizability of axial toroidal moments. As mentioned in chapter 2, such moments are usually considered as not possible in microscopic current distributions. However, no proof of this point of view is known to us.

Also, the results in section 4.3 suggest that electric and toroidal moments are always proportional to each other, disregarding the specific current distribution. Such a connection has nowhere been discussed so far, and more investigation of this observation and the question if the toroidal moments are a independent degree of freedom is needed.

Bibliography

[1] S. Mühlig, C. Menzel, C. Rockstuhl, and F. Lederer, *Multipole analysis of meta-atoms*, Metamaterials **5** (2011), 64–73.

[2] G. N. Afanasiev, *Simplest sources of electromagnetic fields as a tool for testing the reciprocity-like theorems*, J. Phys. D: Appl. Phys. **34** (2001), 539–559.

[3] A. D. Boardman, K. Marinov, N. Zheludev, and V. A. Fedotov, *Dispersion properties of nonradiating configurations: Finite-difference time-domain modeling*, Phys. Rev. E **72** (2005), 036603.

[4] A. J. Devaney and E. Wolf, *Radiating and nonradiating classical current distributions and the fields they generate*, Phys. Rev. D **8** (1973), 1044–1047.

[5] E. Wolf and T. Habashy, *Invisible bodies and uniqueness of the inverse scattering problem*, J. Mod. Opt. **40** (1993), 785–792.

[6] K. Marinov, A. D. Boardman, V. A. Fedotov, and N. Zheludev, *Toroidal metamaterial*, New J. Phys. **9** (2007), 342.

[7] A. E. Miroshnichenko, A. B. Evlyukhin, Y. F. Yu, R. M. Bakker, A. Chipouline, A. I. Kuznetsov, B. Luk'yanchuk, B. N. Chichkov, and Y. S. Kivshar, *Seeing the unseen: observation of an anapole with dielectric nanoparticles*, arXiv:1412.0299 [physics.optics] (2014).

[8] W. Liu, J. Zhang, B. Lei, H. Hu, and A. E. Miroshnichenko, *Invisible nanowires with interferencing electric and toroidal dipoles*, arXiv:1502.02205 [physics.optics] (2015).

[9] J. D. Jackson, *Classical Electrodynamics*, 3rd ed., John Wiley & Sons, New York City, New York, 1998.

[10] R. E. Raab and O. L. de Lange, *Multipole theory in electromagnetism: Classical, Quantum, and Symmetry Aspects, with Applications*, 1st ed., International Series of Monographs on Physics, Clarendon Press, Oxford, 2005.

[11] W. Panofsky and M. Phillips, *Classical electricity and magnetism*, 2nd ed., Addison-Wesley, Reading, 1962.

[12] V. M. Dubovik and L. A. Tosunyan, *Toroidal moments in the physics of electromagnetic and weak interactions*, Sov. J. Part. Nucl. **14** (1983), 504–519.

[13] E. E. Radescu and G. Vaman, *Cartesian multipole expansions and tensorial identities*, Prog. Electromag. Res. B **36** (2012), 89–111.

[14] I. B. Zel'dovich, *Electromagnetic interaction with parity violation*, J. Exptl. Theoret. Phys. **33** (1957), 1531–1533.

[15] V. M. Dubovik and A. A. Cheshkov, *Form factors and multipoles in electromagnetic interactions*, Sov. Phys. JETP **24** (1967), 924–926.

[16] V. M. Dubovik and V. V. Tugushev, *Toroid moments in electrodynamics and solid-state physics*, Phys. Rep. **187** (1990), 145–202.

[17] V. M. Dubovik, L. A. Tosunyan, and V. V. Tugushev, *Axial toroidal moments in electrodynamics and solid-state physics*, Sov. Phys. JETP **63** (1986), 344–351.

[18] J. Van Bladel, *Hierarchy of terms in a multipole expansion*, Electron. Lett. **24** (1988), 492–493.

[19] G. N. Afanasiev and Y. P. Stepanovsky, *The electromagnetic field of elementary time-dependent toroidal sources*, J. Phys. A: Math. Gen. **28** (1995), 4565–4580.

[20] G. N. Afanasiev and V. M. Dubovik, *Some remarkable charge-current configurations*, Phys. Part. Nuclei **29** (1998), 366–391.

[21] V. M. Dubovik, M. A. Martsenyuk, and B. Saha, *Material equations for electromagnetism with toroidal polarizations*, Phys. Rev. E **61** (2000), 7087–7097.

[22] E. E. Radescu and G. Vaman, *Exact calculation of the angular momentum loss, recoil force, and radiation intensity for an arbitrary source in terms of electric, magnetic, and toroid multipoles*, Phys. Rev. E **65** (2002), 046609.

[23] C. Vrejoiu, *Electromagnetic multipoles in Cartesian coordinates*, J. Phys. A: Math. Gen. **35** (2002), 9911–9922.

[24] V. A. Fedotov, K. Marinov, A. D. Boardman, and N. I. Zheludev, *On the aromagnetism and anapole moment of anthracene nanocrystals*, New J. Phys. **9** (2007), 95.

[25] N. Papasimakis, V. A. Fedotov, K. Marinov, and N. I. Zheludev, *Gyrotropy of a meta-molecule: Wire on a torus*, Phys. Rev. Lett. **103** (2009), 093901.

[26] V. A. Fedotov, A. V. Rogacheva, V. Savinov, D. P. Tsai, and N. I. Zheludev, *Resonant transparency and non-trivial non-radiating excitations in toroidal metamaterials*, Sci. Rep. **3** (2013), 2967.

[27] V. Savinov, V. A. Fedotov, and N. I. Zheludev, *Toroidal dipolar excitation and macroscopic electromagnetic properties of metamaterials*, Phys. Rev. B **89** (2014), 205112.

[28] T. Kaelberer, V. A. Fedotov, N. Papasimakis, D. P. Tsai, and N. I. Zheludev, *Toroidal dipolar response in a metamaterial*, Science **330** (2010), 1510–1512.

[29] A. Castellanos, M. Panizo, and J. Rivas, *Magnetostatic multipoles in Cartesian coordinates*, Am. J. Phys. **46** (1978), 1116–1117.

[30] Y. V. Kopaev, *Toroidal ordering in crystals*, Physics – Uspekhi **52** (2009), 1111–1125.

[31] W.-K. Tung, *Group Theory in Physics*, 1st ed., World Scientific Publishing Company, Singapur, 1985.

[32] R. R. Lewis, *Anapole moment of a diatomic polar molecule*, Phys. Rev. A **49** (1994), 3376–3380.

[33] F. J. Lowes and B. Duka, *Magnetic multipole moments (gauss coefficients) and vector potential given by an arbitrary current distribution*, Earth Planets Space **63** (2011), i–vi.

[34] G. N. Afanasiev, M. Nelhiebel, and Y. P. Stepanovsky, *The interaction of magnetizations with an external electromagnetic field and a time-dependent magnetic Aharonov-Bohm effect*, Phys. Scripta **54** (1996), 417–427.

[35] C. Cohen-Tannoudj, J. Dupont-Roc, and G. Grynberg, *Photons and Atoms: Introduction to Quantum Electrodynamics*, John Wiley & Sons, New York City, New York, 1997.

[36] W. W. Hansen, *A new type of expansion in radiation problems*, Phys. Rev. **47** (1935), 139–143.

[37] I. N. Bronstein, K. A. Semendjajew, G. Musiol, and H. Mühlig, *Taschenbuch der Mathematik*, 7th ed., Verlag Harri Deutsch, 2008.

[38] C. G. Gray, *Multipole expansions of electromagnetic fields using Debye potentials*, Am. J. Phys. **46** (1978), 169–179.

[39] C. G. Gray and B. G. Nickel, *Debye potential representation of vector fields*, Am. J. Phys. **46** (1978), 735–736.

[40] http://commons.wikimedia.org/wiki/File:Toroidal_coord.png by User:DaveBurke, retrieved 7 February 2015.

[41] W. M. Elsasser, *Induction effects in terrestrial magnetism*, Phys. Rev. **69** (1946), 106–116.

[42] C. Bohren and D. R. Huffman, *Absorption and Scattering of Light by Small Particles*, 1st ed., John Wiley & Sons, New York City, New York, 1983.

[43] J. Van Bladel, *Electromagnetic Fields*, 2nd ed., John Wiley & Sons, Hoboken, New Jersey, 2007.

[44] B. Blaive and J. Metzger, *Explicit expressions of the nth gradient of 1/r*, J. Math. Phys. **25** (1984), 1721–1724.

[45] M. E. Rose, *Elementary Theory of Angular Momentum*, reprint of the 1st ed., Dover Publications, 1995.

[46] A. Guth, *Lecture notes 9: Traceless symmetric tensor approach to Legendre polynomials and spherical harmonics*, MIT OpenCourseWare, http://ocw.mit.edu/courses/physics/8-07-electromagnetism-ii-fall-2012/lecture-notes/MIT8_07F12_ln9.pdf, 17 October 2012, retrieved 23 January 2015.

[47] D. L. Andrews and W. A. Ghoul, *Irreducible fourth-rank Cartesian tensors*, Phys. Rev. A **25** (1982), 2647–2657.

[48] K. Bansal and G. Sudarshan, *Reduction of Cartesian tensors and its application to stochastic dynamics*, Nuovo Cim. **25** (1962), 1270–1281.

[49] J. A. R. Coope, R. F. Snider, and F. R. McCourt, *Irreducible Cartesian tensors*, J. Chem. Phys. **43** (1965), 2269–2275.

[50] D. L. Andrews and N. P. Blake, *Three-dimensional rotational averages in radiation-molecule interactions: an irreducible Cartesian tensor formulation*, J. Phys. A: Math. Gen. **22** (1989), 49–60.

[51] C. Vrejoiu, *A formula for gauge invariant reduction of electromagnetic multipole tensors*, J. Phys. A: Math. Gen. **38** (2005), L506–L511.

[52] J. Applequist, *Traceless Cartesian tensor forms for spherical harmonic functions: new theorems and applications to electrostatics of dielectric media*, J. Phys. A: Math. Gen. **22** (1989), 4303–4330.

[53] B. Ögüt, N. Talebi, R. Vogelsang, W. Sigle, and P A. van Aken, *Toroidal plasmonic eigenmodes in oligomer nanocavities for the visible*, Nano Lett. **12** (2012), 5239–5244.

[54] I. Dumitriu and C. Vrejoiu, *Some aspects of electromagnetic multipole expansions*, Rom. Rep. Phys. **60** (2008), 423–442.

[55] G. N. Afanasiev, *Vector solutions of the laplace equation and the influence of helicity on the aharonov-bohm scattering*, Phys. Scr. **50** (1994), 225–232.

[56] L. Landau and E. Lifshitz, *The Classical Theory of Fields*, 3rd ed., Course of Theoretical Physics, vol. 2, Pergamon Press, 1971.

[57] V. P. Kazantsev, *Spherical magnetic multipole moments in systems of currents*, Russ. Phys. J. **42** (1999), 916–921.

[58] R. Mehrem, J. T. Londergan, and M. H. Macfarlane, *Analytic expressions for integrals of products of spherical Bessel functions*, J. Phys. A: Math. Gen. **24** (1991), 1435–1453.

[59] N. J. Carron, *On the fields of a torus and the role of the vector potential*, Am. J. Phys. **63** (1995), 717–729.

[60] N. Bleistein and J. K. Cohen, *Nonuniqueness in the inverse source problem in acoustics and electromagnetics*, J. Math. Phys. **18** (1977), 194–201.

[61] A. J. Devaney and E. Wolf, *Multipole expansions and plane wave representations of the electromagnetic field*, J. Math. Phys. **15** (1974), 234–244.

[62] H. Ammari (ed.), *Modeling and Computations in Electromagnetics*, Lecture Notes in Computational Science and Engineering, vol. 59, Springer-Verlag, Berlin Heidelberg New York, 2008.

[63] R. C. Wittmann, *Spherical wave operators and the translation formulas*, IEEE Trans. Antennas. Propag. **36** (1988), 1078–1087.

[64] S. Mühlig, *Towards self-assembled metamaterials*, Ph.D. thesis, Friedrich-Schiller-Universität Jena, 2014.

[65] E. Hebestreit, *Multipole analysis of optical nano-structures*, Masterthesis, Friedrich-Schiller-Universität Jena, September 2013.

[66] J. B. French and Y. Shimamoto, *Theory of multipole radiation*, Phys. Rev. **91** (1953), 898–899.

[67] J.-M. Normand and J. Raynal, *Relations between Cartesian and spherical components of irreducible Cartesian tensors*, J. Phys. A: Math. Gen. **15** (1982), 1437.

[68] J. Chen, J. Ng, Z. Lin, and C. T. Chan, *Optical pulling force*, Nature Photon. **5** (2011), 531–534.

Detailed Calculations

Tensorial Decomposition of Eq. (4.41)

We want to decompose the second order of the dynamic vector potential, Eq. (4.42), which reads as

$$A_i^{(2)}(\vec{r},t) = \frac{\mu_0}{4\pi}\frac{1}{2}\sum_{jk}\int\left(\frac{3r_jr_k - r^2\delta_{jk}}{r^5}\left(j_i(\vec{r}',\tau) + \frac{r}{c}\dot{j}_i(\vec{r}',\tau)\right) + \frac{r_jr_k}{c^2r^3}\ddot{j}_i(\vec{r}',\tau)\right)r_j'r_k'\,\mathrm{d}^3r'$$

into terms with irreducible tensors.

Formulae for Decomposition

The formulae to decompose a reducible tensor of rank 3 into irreducible tensors are taken from [50]. In this section, we use $\hat{T}$ as symbol for a general tensor not to be confused with toroidal multipole moment tensors.

An irreducible tensor of rank 3 can be expressed through the following sum of irreducible tensors:

$$T_{ijk} = T_{ijk}^{(0)} + \sum_{p=1,2,3} T_{ijk}^{(1;p)} + \sum_{p=1,2} T_{ijk}^{(2;p)} + T_{ijk}^{(3)}.$$

These irreducible parts are given through the following formulae:

$$T_{ijk}^{(0)} = \frac{1}{6}\epsilon_{ijk}\epsilon_{abc}T_{abc}$$

$$T_{ijk}^{(1;1)} = \frac{1}{10}(4\delta_{ij}T_{mmk} - \delta_{ik}T_{mmj} - \delta_{jk}T_{mmi})$$

$$T_{ijk}^{(1;2)} = \frac{1}{10}(-\delta_{ij}T_{mkm} + 4\delta_{ik}T_{mjm} - \delta_{jk}T_{mim})$$

$$\hat{T}_{ijk}^{(1;3)} = \frac{1}{10}(-\delta_{ij}T_{kmm} - \delta_{ik}T_{jmm} + 4\delta_{jk}T_{imm})$$

$$T_{ijk}^{(2;1)} = \frac{1}{6}\epsilon_{ijt}(2\epsilon_{mst}T_{msk} + 2\epsilon_{msk}T_{mst} + \epsilon_{mst}T_{kms} + \epsilon_{msk}T_{tms} - 2\delta_{it}\epsilon_{nms}T_{nms})$$

$$T_{ijk}^{(2;2)} = \frac{1}{6}\epsilon_{jkt}(2\epsilon_{mst}T_{ims} + 2\epsilon_{msi}T_{tms} + \epsilon_{mst}T_{msi} + \epsilon_{msi}T_{mst} - 2\delta_{it}\epsilon_{nms}T_{nms})$$

$$T_{ijk}^{(3)} = \frac{1}{6}(T_{ijk} + T_{jik} + T_{ikj} + T_{jki} + T_{kij} + T_{kji})$$
$$- \frac{1}{15}[\delta_{ij}(T_{mmk} + T_{mkm} + T_{kmm}) + \delta_{ik}(T_{mmj} + T_{mjm} + T_{jmm})$$
$$+ \delta_{jk}(T_{mmi} + T_{mim} + T_{imm})]$$

Calculations

We consider for simplicity only the tensor

$$T_{ijk} = \int j_i r_j r_k \, \mathrm{d}^3 r \,.$$

We get:

$$T_{ijk}^{(0)} = \frac{1}{6}\int \epsilon_{ijk}(j_a r_b r_c - j_a r_c r_b - j_c r_b r_a + j_c r_a r_b - j_b r_a r_c + j_b r_c r_a)\,\mathrm{d}^3 r = 0$$

$$T_{ijk}^{(1;1)} = \frac{1}{10}\int (4\delta_{ij}j_m r_m r_k - \delta_{ik}j_m r_m r_j - \delta_{jk}j_m r_m r_i)\,\mathrm{d}^3 r$$
$$= \frac{1}{10}\int (4\delta_{ij}\vec{r}\cdot\vec{j}\,r_k - \delta_{ik}\vec{r}\cdot\vec{j}\,r_j - \delta_{jk}\vec{r}\cdot\vec{j}\,r_i)\,\mathrm{d}^3 r$$

$$T_{ijk}^{(1;2)} = \frac{1}{10}\int (-\delta_{ij}j_m r_k r_m + 4\delta_{ik}j_m r_j r_m - \delta_{jk}j_m r_i r_m)\,\mathrm{d}^3 r$$
$$= \frac{1}{10}\int (-\delta_{ij}\vec{r}\cdot\vec{j}\,r_k + 4\delta_{ik}\vec{r}\cdot\vec{j}\,r_j - \delta_{jk}\vec{r}\cdot\vec{j}\,r_i)\,\mathrm{d}^3 r$$

$$T_{ijk}^{(1;3)} = \frac{1}{10}\int (-\delta_{ij}j_k r_m r_m - \delta_{ik}j_j r_m r_m + 4\delta_{jk}j_i r_m r_m)\,\mathrm{d}^3 r$$
$$= \frac{1}{10}\int (-\delta_{ij}j_k r^2 - \delta_{ik}j_j r^2 + 4\delta_{jk}j_i r^2)\,\mathrm{d}^3 r$$

$$\sum_{p=1,2,3} T_{ijk}^{(1;p)} = \frac{1}{10}\int (3\delta_{ij}\vec{r}\cdot\vec{j}\,r_k + 3\delta_{ik}\vec{r}\cdot\vec{j}\,r_j - 2\delta_{jk}\vec{r}\cdot\vec{j}\,r_i - \delta_{ij}j_k r^2 - \delta_{ik}j_j r^2 + 4\delta_{jk}j_i r^2)\,\mathrm{d}^3 r$$
$$= \frac{1}{10}\int [\delta_{ij}(3\vec{r}\cdot\vec{j}\,r_k - j_k r^2) + \delta_{ik}(3\vec{r}\cdot\vec{j}\,r_j - j_j r^2) + 2\delta_{jk}(2j_i r^2 - \vec{r}\cdot\vec{j}\,r_i)]\,\mathrm{d}^3 r$$
$$= \delta_{ij}\left(t_k + \frac{1}{10}r_{\vec{p},k}^{(2)}\right) + \delta_{ik}\left(t_j + \frac{1}{10}r_{\vec{p},j}^{(2)}\right) - 2\delta_{jk}t_i$$

$$T_{ijk}^{(2;1)} = \frac{1}{6}\epsilon_{ijt} \int (2\epsilon_{mst}j_m r_s r_k + 2\epsilon_{msk}j_m r_s r_t + \epsilon_{mst}j_k r_m r_s + \epsilon_{msk}j_t r_m r_s - 2\delta_{kt}\epsilon_{nms}j_n r_m r_s)\, \mathrm{d}^3 r$$

$$= \frac{1}{6}\epsilon_{ijt} \int [2(\vec{j} \times \vec{r})_t r_k + 2(\vec{j} \times \vec{r})_k r_t]\, \mathrm{d}^3 r$$

$$= -\epsilon_{ijt}Q_{kt}^{(m)}$$

$$T_{ijk}^{(2;2)} = \frac{1}{6}\epsilon_{jkt} \int (2\epsilon_{mst}j_i r_m r_s + 2\epsilon_{msi}j_t r_m r_s + \epsilon_{mst}j_m r_s r_i + \epsilon_{msi}j_m r_s r_t - 2\delta_{it}\epsilon_{nms}j_n r_m r_s)\, \mathrm{d}^3 r$$

$$= \frac{1}{6}\epsilon_{jkt} \int [(\vec{j} \times \vec{r})_t r_i + (\vec{j} \times \vec{r})_i r_t]\, \mathrm{d}^3 r$$

$$= -\frac{1}{2}\epsilon_{jkt}Q_{it}^{(m)}$$

There seems to be a mistake in the paper where we took the formulae from, because the two contributions for the magnetic dipole moment should have the same prefactor [16].

$$T_{ijk}^{(3)} = \int \left[\frac{1}{6} (j_i r_j r_k + j_i r_k r_j + j_k r_j r_i + j_k r_i r_j + j_j r_i r_k + j_j r_k r_i) \right.$$

$$- \frac{1}{15}(\delta_{ij}(j_m r_m r_k + j_m r_k r_m + j_k r_m r_m) + \delta_{ik}(j_m r_m r_j + j_m r_j r_m + j_j r_m r_m)$$

$$\left. + \delta_{jk}(j_m r_m r_i + j_m r_i r_m + j_i r_m r_m)) \right] \mathrm{d}^3 r$$

$$= \frac{1}{3} \int \left((j_i r_j r_k + j_k r_i r_j + j_j r_i r_k) - \frac{1}{5}(\delta_{ij}(j_k r^2 + 2r_k \vec{r} \cdot \vec{j}) \right.$$

$$\left. + \delta_{ki}(j_j r^2 + 2r_j \vec{r} \cdot \vec{j}) + \delta_{jk}(j_i r^2 + 2r_i \vec{r} \cdot \vec{j})) \right) \mathrm{d}^3 r$$

$$= \frac{1}{3} \int \left(r_i r_j r_k - \frac{1}{5}r^2(\delta_{ij}r_k + \delta_{ki}r_j + \delta_{jk}r_i) \right) \dot{\rho}\, \mathrm{d}^3 r$$

$$= \frac{1}{3}\dot{O}_{ijk}^{(e)}$$

The quantities, which have been defined here, are explained in the main text. The sum of all these irreducible tensors is

$$T_{ijk} = \frac{1}{3}\dot{O}_{ijk}^{(e)} - \frac{1}{3}(2\epsilon_{ijt}Q_{kt}^{(m)} + \epsilon_{jkt}Q_{it}^{(m)}) + \delta_{ij}t_k + \delta_{ik}t_j - 2\delta_{jk}t_i + \frac{1}{10}(\delta_{ij}r_{\vec{p},k}^{(2)} + \delta_{ik}r_{\vec{p},j}^{(2)}).$$

Plugging this sum into the formula for the second order of the vector potential yields

$$\vec{A}^{(2)}(\vec{r},t) = \frac{\mu_0}{4\pi} \frac{1}{2r^5} \left[\vec{r}^T \dot{\hat{O}}^{(e)} \cdot \vec{r} + \frac{r}{c} \vec{r}^T \ddot{\hat{O}}^{(e)} \cdot \vec{r} + \frac{r^2}{3c^2} \vec{r}^T \dddot{\hat{O}}^{(e)} \cdot \vec{r} - 2\vec{r} \times (\hat{Q}^{(m)} \cdot \vec{r}) \right.$$

$$- \frac{2r}{c} \vec{r} \times (\dot{\hat{Q}}^{(m)} \cdot \vec{r}) - \frac{2r^2}{3c^2} \vec{r} \times (\ddot{\hat{Q}}^{(m)} \cdot \vec{r}) + 6\vec{r}(\vec{r} \cdot \vec{t}) - 2r^2 \vec{t} + \frac{6r}{c}\vec{r}(\vec{r} \cdot \dot{\vec{t}}) - \frac{2r^3}{c} \ddot{\vec{t}}$$

$$\left. + \frac{2r^2}{c^2} \vec{r}(\vec{r} \cdot \ddot{\vec{t}}) + \frac{9}{5}\vec{r}(\vec{r} \cdot \vec{r}_p^{(2)}) + \frac{3r}{5c}\vec{r}(\vec{r} \cdot \dot{\vec{r}}_p^{(2)}) + \frac{3r^2}{5c^2}\vec{r}(\vec{r} \cdot \ddot{\vec{r}}_p^{(2)}) - \frac{r^2}{5}\vec{r}_p^{(2)} - \frac{r^3}{5c}\dot{\vec{r}}_p^{(2)} \right] .$$

This is Eq. (4.42).

Proof of Eq. (4.27)

We want to proof Eq. (4.27) using mathematical induction.

The statement to prove is

$$\nabla_i \left[\prod_{m=1}^{n} r_{i_m} j_i \right] = n j_{i_1} \prod_{m=2}^{n} r_{i_m} - \prod_{m=1}^{n} r_{i_m} \dot{\rho} + \sum_{k=1}^{n} \epsilon_{i_1 i_k l} (\vec{r} \times \vec{j})_l \prod_{\substack{m=2 \\ m \neq k}}^{n} r_{i_m} .$$

Base case: $n = 2$:

$$\nabla_i [r_{i_1} r_{i_2} j_i] = 2j_{i_1} r_{i_2} + r_{i_1} r_{i_2} \nabla_i \vec{j}_i + \epsilon_{i_1 i_1 l}(\vec{r} \times \vec{j})_l \prod_{\substack{m=2 \\ m \neq 1}}^{2} r_{i_m} + \epsilon_{i_1 i_2 l}(\vec{r} \times \vec{j})_l \prod_{\substack{m=2 \\ m \neq 2}}^{2} r_{i_m}$$

$$= 2j_{i_1} r_{i_2} - r_{i_1} r_{i_2} \dot{\rho} + \epsilon_{i_1 i_2 l}(\vec{r} \times \vec{j})_l$$

We used that for the empty product it holds by definition

$$\prod_{\substack{m=2 \\ m \neq 2}}^{2} r_{i_m} = 1 .$$

Inductive step:

$$\nabla_i \left[\prod_{m=1}^{n+1} r_{i_m} j_i \right] = \nabla_i \left[\prod_{m=1}^{n} r_{i_m} j_i r_{n+1} \right]$$

$$= \nabla_i \left[\prod_{m=1}^{n} r_{i_m} j_i \right] r_{n+1} + \left[\prod_{m=1}^{n} r_{i_m} j_i \right] \nabla_i r_{n+1}$$

$$= \left[n j_{i_1} \prod_{m=2}^{n} r_{i_m} - \prod_{m=1}^{n} r_{i_m} \dot{\rho} + \sum_{k=1}^{n} \epsilon_{i_1 i_k l}(\vec{r} \times \vec{j})_l \prod_{\substack{m=2 \\ m \neq k}}^{n} r_{i_m} \right] r_{i_{n+1}} + j_{i_{n+1}} \prod_{m=1}^{n} r_{i_m}$$

Using [10, p. 211]

$$j_j r_i - j_i r_j = \epsilon_{ijk}(\vec{r} \times \vec{j})_k ,$$

it follows for arbitrary i, j

$$j_{i_{n+1}} \prod_{m=1}^{n} r_{i_m} = j_{i_{n+1}} r_{i_1} \prod_{m=2}^{n} r_{i_m} = \left[j_{i_1} r_{i_{n+1}} + \epsilon_{i_1 i_{n+1} p} (\vec{r} \times \vec{j})_p \right] \prod_{m=2}^{n} r_{i_m} ,$$

and thus as last step of the proof

$$\nabla_i \left[\prod_{m=1}^{n+1} r_{i_m} j_i \right] = \left[n j_{i_1} \prod_{m=2}^{n} r_{i_m} - \prod_{m=1}^{n} r_{i_m} \dot{\rho} + \sum_{k=1}^{n} \epsilon_{i_1 i_k l} (\vec{r} \times \vec{j})_l \prod_{\substack{m=2 \\ m \neq k}}^{n} r_{i_m} \right] r_{i_{n+1}} + j_{i_{n+1}} \prod_{m=1}^{n} r_{i_m}$$

$$= \left[n j_{i_1} \prod_{m=2}^{n} r_{i_m} - \prod_{m=1}^{n} r_{i_m} \dot{\rho} + \sum_{k=1}^{n} \epsilon_{i_1 i_k l} (\vec{r} \times \vec{j})_l \prod_{\substack{m=2 \\ m \neq k}}^{n} r_{i_m} \right] r_{i_{n+1}}$$

$$+ \left[j_{i_1} r_{i_{n+1}} + \epsilon_{i_1 i_{n+1} p} (\vec{r} \times \vec{j})_p \right] \prod_{m=2}^{n} r_{i_m}$$

$$= (n+1) j_{i_1} \prod_{m=2}^{n+1} r_{i_m} - \prod_{m=1}^{n+1} r_{i_m} \dot{\rho} + \sum_{k=1}^{n+1} \epsilon_{i_1 i_k l} (\vec{r} \times \vec{j})_l \prod_{\substack{m=2 \\ m \neq k}}^{n+1} r_{i_m} .$$

Proof of Eq. (5.16)

The transverse part of $\vec{j}$ can, like the electric field in section 5.1, be represented as

$$\vec{j}_\perp = \sum_{jm} [a_{jm} \vec{M}_{jm} + b_{jm} \vec{N}_{jm}]$$

with

$$\vec{\nabla} \times \vec{M} = -\mathrm{i} k \vec{N}$$
$$\vec{\nabla} \times \vec{N} = k \vec{M}$$

and thus

$$\vec{\nabla} \times \vec{\nabla} \times \vec{M} = k^2 \vec{M}$$
$$\vec{\nabla} \times \vec{\nabla} \times \vec{N} = k^2 \vec{N}$$

and

$$\vec{\nabla} \times \vec{\nabla} \times \vec{j}_\perp = k^2 \vec{j}_\perp$$

This yields

$$[\Delta + k^2] \vec{E}_\perp = -\mathrm{i} \omega \mu_0 \frac{1}{k^2} \vec{\nabla} \times \vec{\nabla} \times \vec{j}_\perp - \frac{\mathrm{i}}{\omega \epsilon_0} \vec{\nabla} \times \vec{\nabla} \times \vec{j}_\parallel = -\frac{\mathrm{i}}{\omega \epsilon_0} \vec{\nabla} \times \vec{\nabla} \times \vec{j} .$$

Tabulated Functions

Spherical Harmonics

The spherical harmonics Y_{lm} for indices up to $l = 2$ are:

$l = 0$:

$$Y_{00} = \frac{1}{\sqrt{4\pi}}$$

$l = 1$:

$$Y_{10} = \frac{1}{2}\sqrt{\frac{3}{\pi}}\cos\theta$$

$$Y_{11} = -\frac{1}{2}\sqrt{\frac{3}{2\pi}}e^{i\phi}\sin\theta$$

$$Y_{1-1} = \frac{1}{2}\sqrt{\frac{3}{2\pi}}e^{-i\phi}\sin\theta$$

$l = 2$:

$$Y_{20} = \frac{1}{4}\sqrt{\frac{5}{\pi}}\left(3\cos^2\theta - 1\right)$$

$$Y_{21} = -\frac{1}{2}\sqrt{\frac{15}{2\pi}}e^{i\phi}\sin\theta\cos\theta$$

$$Y_{2-1} = \frac{1}{2}\sqrt{\frac{15}{2\pi}}e^{-i\phi}\sin\theta\cos\theta$$

$$Y_{22} = \frac{1}{4}\sqrt{\frac{15}{2\pi}}e^{2i\phi}\sin^2\theta$$

$$Y_{2-2} = \frac{1}{4}\sqrt{\frac{15}{2\pi}}e^{-2i\phi}\sin^2\theta$$

Spherical Bessel and Hankel functions

The spherical Bessel functions, $j_l(kr)$, and spherical Hankel functions of first kind, $h_l^{(1)}(kr)$, up to index $l = 2$ are:

$l = 0$:

$$j_0(kr) = \frac{\sin(kr)}{kr}$$

$l = 1$:

$$j_1(kr) = \frac{\sin(kr)}{k^2 r^2} - \frac{\cos(kr)}{kr}$$

$l = 2$:

$$j_2(kr) = \frac{(3 - k^2 r^2)\sin(kr)}{k^3 r^3} - \frac{3\cos(kr)}{k^2 r^2}$$

$l = 0$:

$$h_0^{(1)}(kr) = -\frac{ie^{ikr}}{kr}$$

$l = 1$:

$$h_1^{(1)}(kr) = \frac{e^{ikr}(-kr - i)}{k^2 r^2}$$

$l = 2$:

$$h_2^{(1)}(kr) = \frac{ie^{ikr}(k^2 r^2 + 3ikr - 3)}{k^3 r^3}$$

Vector Spherical Harmonics

The Cartesian components of the vector spherical harmonics are [22]

$$(\vec{Y}_{llm})_x = -\frac{c_1}{\sqrt{2}}Y_{lm-1} + \frac{c_3}{\sqrt{2}}Y_{lm+1},$$

$$(\vec{Y}_{llm})_y = -\frac{ic_1}{\sqrt{2}}Y_{lm-1} - \frac{ic_3}{\sqrt{2}}Y_{lm+1},$$

$$(\vec{Y}_{llm})_z = c_2 Y_{lm},$$

where

$$c_1 = -\frac{\sqrt{(l+m)(l-m+1)}}{\sqrt{l(2l+2)}},$$

$$c_2 = \frac{m}{\sqrt{l(l+1)}},$$

$$c_3 = \frac{\sqrt{(l-m)(l+m+1)}}{\sqrt{l(2l+2)}},$$

$$(\vec{Y}_{ll-m})_x = -\frac{c_1}{\sqrt{2}}Y_{l-1m-1} + \frac{c_3}{\sqrt{2}}Y_{l-1m+1},$$

$$(\vec{Y}_{ll-m})_y = -\frac{ic_1}{\sqrt{2}}Y_{l-1m-1} - \frac{ic_3}{\sqrt{2}}Y_{l-1m+1},$$

$$(\vec{Y}_{llm})_z = c_2 Y_{l-1m},$$

where

$$c_1 = \frac{\sqrt{(l+m-1)(l+m)}}{\sqrt{l(2l-1)}},$$

$$c_2 = \frac{(l-m)(l+m)}{\sqrt{2l(2l+1)}},$$

$$c_3 = \frac{\sqrt{(l-m-1)(l-m)}}{\sqrt{2l(2l-1)}},$$

$$(\vec{Y}_{ll+m})_x = -\frac{c_1}{\sqrt{2}}Y_{l+1m-1} + \frac{c_3}{\sqrt{2}}Y_{l+1m+1},$$

$$(\vec{Y}_{ll+m})_y = -\frac{ic_1}{\sqrt{2}}Y_{l+1m-1} - \frac{ic_3}{\sqrt{2}}Y_{l+1m+1},$$

$$(\vec{Y}_{llm})_z = c_2 Y_{l+1m},$$

where

$$c_1 = \frac{\sqrt{(l-m+1)(l-m+2)}}{\sqrt{(2l+2)(2l+3)}} \; ,$$

$$c_2 = \frac{(l+m+1)(l-m+1)}{\sqrt{(l+1)(2l+3)}} \; ,$$

$$c_3 = \frac{\sqrt{(l+m+1)(l+m+2)}}{\sqrt{(2l+2)(2l+3)}} \; .$$

Printed in the United States
By Bookmasters